MEGATECH

MEGATECH

TECHNOLOGY IN 2050

edited by

DANIEL FRANKLIN

PUBLICAFFAIRS
New York

Contents

Contributors

Ryan Avent is economics columnist at *The Economist*. He is the author of *The Wealth of Humans: Work and its Absence in the Twenty-First Century*.

Geoffrey Carr is *The Economist*'s science and technology editor.

Tim Cross is science correspondent at *The Economist*.

Robert Carlson is a scientist, entrepreneur and the author of *Biology is Technology: The Promise, Peril, and New Business of Engineering Life*. He is a managing director of Bioeconomy Capital, a biotech investment house, and a principal at Biodesic, a strategy, engineering and security consulting firm.

Kenneth Cukier is senior editor for digital products at *The Economist*. He is the co-author of *Big Data: A Revolution that Transforms How We Work, Live and Think*.

Gianrico Farrugia is chief executive of the Mayo Clinic at Jacksonville, Florida. A professor of medicine and physiology, he specialises in gastroenterology and is a pioneer of individualised medicine.

Luciano Floridi is professor of philosophy and ethics of information at the University of Oxford. His most recent book is *The Fourth Revolution: How the Infosphere is Reshaping Human Reality*.

Melinda Gates is co-founder and co-chair of the Bill & Melinda Gates Foundation.

Lynda Gratton is professor of management practice at London Business School. Her books on the impact of a changing world on work include *The Shift* and *The Key*. Her latest book, with co-author Andrew Scott, is *The 100-Year Life: Living and Working in an Age of Longevity*.

Nancy Kress is a multiple-award-winning American science-fiction writer, whose works include *After the Fall, Before the Fall, During the Fall* and *Yesterday's Kin*.

Paul Markillie is innovation editor at *The Economist*. He was previously editor of the newspaper's Technology Quarterly and author of a special report on the future of manufacturing, *The third industrial revolution*.

Leo Mirani is *The Economist*'s news editor. Previously he reported on technology for *Quartz*.

Oliver Morton is *The Economist*'s briefings editor and a seasoned science writer. His latest book is *The Planet Remade: How Geoengineering Could Change the World*.

Alastair Reynolds is a British science-fiction writer, whose works include the *Poseidon's Children* and *Revelation Space* series of novels.

Anne Schukat is a freelance journalist who has been a frequent contributor to *The Economist* on science and technology.

Tom Standage is deputy editor and head of digital strategy at *The Economist*. He is one of the paper's leading writers on technology. His books include *The Victorian Internet* and *Writing on the Wall: Social Media – The First 2,000 Years*.

Benjamin Sutherland is a regular contributor to *The Economist*. He is the author of *Modern Warfare, Intelligence and Deterrence: The Technologies that are Transforming Them*.

Frank Wilczek is Herman Feshbach professor of physics at the Massachusetts Institute of Technology. He was awarded the Nobel prize in physics in 2004. His books include *A Beautiful Question: Finding Nature's Deep Design.*

Ann Winblad is a founding partner of Hummer Winblad Venture Partners, based in San Francisco, and for over 30 years has been a leading entrepreneur and investor in the software industry.

Adrian Wooldridge is *The Economist*'s management editor and Schumpeter columnist. His most recent book is *The Great Disruption: How Business is Coping with Turbulent Times.*

The editor

Daniel Franklin is executive editor of *The Economist* and editor of its annual publication on the year ahead, *The World in...* He co-edited *Megachange: The World in 2050.*

Introduction: meet megatech

Daniel Franklin

THIS BOOK IS BASED ON the idea that it can be useful to consider the long view. Setting our sights on 2050 is an invitation to identify the fundamental forces likely to shape the world between now and then. This volume's predecessor, *Megachange: The World in 2050*, published in 2012, provided an overview of such trends, from demography and religion to the economy and culture. Here the focus is narrower – on technology alone – but *Megatech* still ranges widely. For technology influences pretty much everything.

Clearly, it is impossible to know for sure what the technologies of 2050 will be, just as, 30 years ago, nobody could have envisaged today's world of Apple, Amazon, Facebook and Google. However, it is interesting and mind-stretching to make educated guesses. To do so, *Megatech* draws on the expertise of scientists, entrepreneurs, academics and sci-fi writers, as well as journalists from *The Economist*. The result is a rich variety of perspectives on how technology will evolve and affect us in the decades ahead.

Tools and platforms

We start with the basics. The first six chapters, in Part 1, address fundamental questions about the future of technology and what is likely to drive or constrain change. Where should we look for signs of what lies ahead? What will advances in science – physics and biology, in particular – make possible, and where might technology bump up against limits? How do investors spot emerging technologies and where are they putting their money now? Will change really be as fast and dramatic as is commonly supposed, or will it pale by comparison with the technology revolution of the last century?

To set about predicting the tech future, it helps to have a toolkit. Tom Standage provides one. He suggests that clues can be found in the patterns of the past, in the "edge cases" of the present and in the "imagined futures" of science fiction. He then tests these tools on four promising areas: virtual reality, self-driving cars, private spaceflight and gene editing. These examples suggest that a fertile period for discovery lies ahead (two science-fiction writers have dubbed the coming period of rapid change "the Accelerando"), which could echo the scientific revolution of the mid-17th century.

Advances in science make an Accelerando seem plausible. In his masterful overview of the state of fundamental physics (many readers will find themselves wishing the subject had been explained so clearly when they were at school), Frank Wilczek makes a striking claim:

> We have, today, accurate, complete equations adequate to provide the foundation of nuclear physics, materials science, chemistry and all plausible forms of engineering.

As a result, calculation can increasingly replace experimentation in developing technology, allowing far faster progress. This offers "brilliant opportunities for creativity in the service of human ends", and opens up "inspiring prospects for achieving new levels of material wealth and spiritual enrichment". Yet it also presents profound perils (or "failure modes"), the most worrying being nuclear war, ecological collapse and artificial-intelligence warfare.

If physics has reached a creative level of maturity, biology bubbles with youthful excitement. In the decades to 2050 we will learn how all the parts and systems underlying life fit together, predicts Robert Carlson. The sorts of things we can expect in the years to come include our brains being plugged into the internet and our used body parts being swapped out for new ones. All this will raise searching ethical questions. Meanwhile whole industries (from food to pharmaceuticals) will be transformed by bioengineering, as it becomes a platform to "build just about anything we see in nature", and much more besides.

Behind biotechnology's formidable potential lies the "hyper-exponential" increase in the productivity of DNA sequencing. A decade ago *The Economist* called this soaring efficiency "Carlson's curve", comparing it to a similar relentless improvement in microchips, known as "Moore's law", which has driven digital development. But Moore's law is running out of steam. Does this mean that the massive demands for computing power – needed to do many of the wondrous things described elsewhere in this book – will bump up against physical limits in future? The short answer, according to Tim Cross, is: probably not. Other technologies will come to the rescue. Progress will be less regular and predictable without the "master metronome" of Moore's law, but a combination of 3D chips, quantum computers and having more of the processing work done in big data centres (hidden away in the "cloud") will enable the computing revolution to continue.

It will take the form of a succession of technology "waves", judging from the experience of recent decades. Half a dozen such waves have rolled in since the 1950s, from the early mainframes to today's smart machines and the "internet of things". In each wave a crowd of companies emerge, but only a few make it to the shore. And each successive wave is stronger than the last, boosted by the force of its predecessors. Silicon Valley investors are already riding the newest (seventh) wave, still in the early stages of formation: it carries artificial-intelligence (AI) companies. Early-stage venture capitalists began investing in AI around 2010, and billions of dollars are now pouring into firms developing AI software tools and applications. Ann Winblad reckons that "a rapid, virtuous and competitive cycle of innovation has picked up invisible momentum as the seventh wave builds". Its force will be felt in the decades ahead.

Yet how big will the impact of AI and other new technologies really be? An American economist, Robert Gordon, is among those who argue that the digital revolution, however impressive, has relatively limited transformative potential when compared with the great innovations of the second half of the 19th century. Electricity, cars, indoor plumbing and modern medicine powered a century of rapid productivity growth; today, despite the spread of the internet, smartphones, apps and bots, productivity and pay are rising at disappointingly slow rates. If anything, technology is contributing to inequality and fuelling frustration. As Ryan Avent explains, however, there is a strong case for greater optimism about the decades ahead. Learning to make the most of new technologies takes time; that was equally true of electrification (indeed, the pattern of labour-productivity growth in the information-technology era is remarkably similar so far to that seen in the electrification age). As with previous tech-driven spurts in economic growth, tomorrow's advances will come in new ways that are hard to imagine now. That is not to say the rapid change ahead will be easy to cope with. On the contrary, it will be difficult and disruptive – something the later chapters of this book grapple with in more depth.

Industrial revolutions

First, though, we look in Part 2 at the transformation technology will bring about in a number of critical industries. Of these, none is more important than farming. How do you feed a planet of nearly 10 billion people by 2050? Comfortably, argues Geoffrey Carr, provided consumers accept the sorts of food-production techniques that will become possible in the not-so-distant future. Such techniques include the application to plants of precise gene-editing tools that could, for example, turbo-boost photosynthesis to make crops grow faster and dramatically improve yields. Urban fish farms could in effect bring the ocean inland and make fish the dominant source of animal protein. Unless, that is, it is outdone by mass production of animal products – steak, milk and eggs without shells should be on the menu – grown from cell culture, without any actual animals.

If there is no reason for the world to go hungry, there is every reason to expect it to be healthier. Health care has in the past been relatively slow to adopt new technology. Yet the pace of change is accelerating. Disruption will come from many areas, including AI, big data and ever-cheaper genome sequencing. The field will start to look very different. New apps and ever more sophisticated AI tools will do jobs once performed by doctors; "targeted therapies", aimed at specific molecules or cells, will dominate drug development; and whole new sub-industries will emerge, around regenerative medicine, for example, and data aggregation. But in one key respect, suggests Gianrico Farrugia, the result will look rather familiar: health care will more closely resemble other industries, with the patient seen as the customer.

The energy industry needs (for the planet's sake) to look less familiar in future, moving away from reliance on fuels that contribute to climate change. Anne Schukat expects to see a great shift away from fossil fuels in the next decades, and a rapid rise of renewable sources of energy, especially solar and wind power, the costs of which are tumbling. Big improvements in battery technology will help: "distributed" storage of energy, in homes as well as businesses, will spread. The world used to worry about energy scarcity. But with the rise of renewables, and with "fracking" technology unlocking stores of oil and gas, the prospect instead is of energy abundance.

New materials will help to make manufacturing a lot more energy-efficient, too. As Paul Markillie points out, the way the BMW i3 electric car is "knitted" together using carbon fibre gives a glimpse of the future: production of the i3 uses 50% less energy and 70% less water than would be the case in a factory using traditional processes and materials. This is part of a materials revolution that includes, beyond carbon fibre, possibilities such as "smart" materials capable of remembering their shape and assembling themselves into components, and molecular-level manipulation to create bespoke substances and to change the way materials respond to light, electricity, water and heat. Clever new materials will also help the spread of "additive manufacturing", popularly known as 3D printing. As materials and processes become critical elements of firms' competitive advantage, a lot of the manufacturing that went offshore will come home, to be closer to customers.

New materials will also have military applications – giving soldiers lighter and more flexible armour, for example. And other technologies, including laser guns and military robots, will be on the march. The US still leads in the making of defence kit, but potential rivals are catching up. By mid-century, says Benjamin Sutherland, the West's monopoly on precision warfare may well be long gone. One hope for the West is that it can keep an edge thanks to a cultural advantage: the freedom of thought that may allow its soldiers to make more effective use of intelligence delivered via smart devices, such as "augmented reality" (AR) displays.

Yet such technologies will be spreading far and wide, anyway. They will, predicts Leo Mirani, change human behaviour even more than the advent of smartphones and the web has done. He describes a 2050 world in which AR glasses have replaced smartphones, conversations with people who speak other languages will be simultaneously translated, and you need never forget a name as everything you know about a person will appear as you talk to them. He imagines technology moving ever closer to our bodies, and even inside them. As this happens, concerns about the amount of data collected about us, and what the companies that gather the information might do with it, will intensify.

Nigh society

By now it should be obvious that the social and policy implications of the technologies on the horizon are huge. This is the focus of the contributions in Part 3, starting with a mind-clearing look at an area that has drawn sombre warnings from the likes of Stephen Hawking and Elon Musk: artificial intelligence. Might ultra-intelligent machines pose an existential threat to the human race? Luciano Floridi argues that the machines won't be the problem, but the humans who create the environment for them could be.

Despite the pitfalls there is enormous scope for progress. In a data-driven world, Kenneth Cukier points out, things that are currently hard to do will become easier, things that are expensive will become cheaper and things that are scarce will become more abundant. So doctors will use big-data systems to help them make better decisions, teachers will use data to tailor the pace of instruction to individual students, lawyers will be able to find relevant evidence and precedents faster and more cheaply. These and other professions won't be swept away by technology – we may even want more doctors, teachers and lawyers, not fewer – but they will need to change their ways and learn new skills.

If the coming digital dynamics mean upheaval in developed economies, this is not the only way that technology can bring dramatic change. Just as significant, if not more so, is the spread to poorer countries of what is already common in rich ones. Melinda Gates imagines a world in which every woman has a smartphone. The effect – from health to farming and banking – would be transformational. And by 2050 it is surely possible.

This is one example of how technology could reduce inequality in the world. Adrian Wooldridge suggests others – making the case that, having been responsible for much of the rise in inequality in recent years, technology could in future help to reverse it. For instance, it can help to detect and select talented youngsters regardless of social background, giving those who might otherwise languish a chance to shine. Whether in education and health, in tackling corruption or in making poorer neighbourhoods safer, it offers a powerful tool for policymakers.

With so much disruption on the horizon, worries are growing about what it will mean for the world of work. Will machines hollow out industries or create more employment? And will new jobs come fast enough to avoid mass misery for those displaced? Lynda Gratton identifies the questions that businesses and policymakers should be asking as they contemplate an uncertain future for work, and concludes that successful organisations will have adaptability at their core. This means creativity in designing career ladders, imagination in nurturing talent, flexibility in training and a fresh attitude to machines-as-partners: "what great feats can be accomplished by workers with their robotic co-workers?"

Imagination, remember, was (in the form of science fiction) part of Tom Standage's toolkit for looking at the tech future. So we include some "imagined futures" in these pages, inviting two sci-fi writers to contribute short stories set in 2050. Alastair Reynolds and Nancy Kress responded splendidly, bringing technological possibilities to life, along with the moral issues they raise. Their works of fiction feel remarkably real.

Risky bigness

Three strands run through *Megatech*, intertwining in intriguing ways from beginning to end. The first reflects the quality suggested by the title: a sense of bigness. The possibilities opened up by the technologies envisaged for 2050 are huge. It is hard not to feel excited about the extraordinary advances that will be within reach. Here is the tantalising prospect of a world where services are delivered faster, cheaper and better; where access to them is widened, reducing inequality; where food is abundant, energy cleaner and transport safer; where people are healthier and have more opportunities.

Yet a lot could go wrong, as Oliver Morton emphasises in his thought-provoking concluding chapter. There will be unintended consequences, potentially dangerous disruption and misuse of technology's power. A wariness of the risks ahead is the second thread weaving its way through these chapters: megatech could become negatech. Frank Wilczek warns of "failure modes", others caution that policymakers will struggle to keep up with the questions posed by what technology makes possible, industry after industry faces upheaval – with all that implies for jobs and the lives of those that work in them. Luciano Floridi puts the pace of change pithily in perspective:

> *The agricultural revolution took millennia to exert its full impact on society, the industrial revolution took centuries, but the digital one only a few decades. No wonder we feel confused and wrong-footed.*

An Accelerando is hard as well as exciting.

Interlaced with the idea of bigness and risk, however, is a third theme: a recurrent notion that there is nothing inevitable about what lies ahead. The impact of technology is only partly a matter of the innovations of scientists, geeks and entrepreneurs. The outcome by 2050 will also be shaped by the decisions of governments, the strategies of companies and the choices of individuals. It is up to all of us to make the most of megatech.

PART 1

The fundamentals

1

A toolkit for predicting the future

Tom Standage

To see what lies ahead in technology, it helps to look in three places: the past, the present and the imagined futures of science fiction

A NEW NETWORKING TECHNOLOGY revolutionises long-distance communication, making it cheaper and more convenient than ever before. It is enthusiastically embraced by businesses, causing a speculative boom. The new technology is relentlessly hyped by its advocates and mocked by its detractors. It makes possible new business models and new forms of crime. Governments struggle to prevent the use of cryptography, demanding access to all messages. People make friends and fall in love online. Some say the new technology will lead to world peace, as communication erases borders and unites humanity. It sounds like the story of the internet in the 1990s. But this is in fact the story of the electric telegraph in the mid-19th century, which was known as the "great highway of thought".

The striking parallels between these two technologies, one modern, one 150 years old, are entertaining – but they can also be useful. The study of history is one of three tools that can be used to predict the future of technology, or at least make slightly more educated guesses about it.

History lessons

Historical analogies of this kind, across years, decades or even centuries, make it possible to foresee the social and cultural impact of new inventions, put hype and scepticism into perspective, provide clues about how a technology might evolve in future, and provide a reminder that problems blamed on new technologies are often the result of human nature. There were, for example, instances of what we would now call "cybercrime" on the mechanical telegraph networks built in the age of Napoleon. "It is a well-known fact that no other section of the population avail themselves more readily and speedily of the latest triumphs of science than the criminal class," in the words of one law-enforcement official. Those words could have been spoken today, but were in fact spoken by a Chicago policeman in 1888.

Such analogies are never perfect, of course, and history never repeats itself exactly. But analogies do not have to be perfect to be informative. Look closely, and there are many repeating patterns in the history of technology, on both short and long timescales.

New inventions often provoke concerns that they will destroy privacy; the first Kodak camera caused a panic over surreptitious public photography in the 1880s, much as Google Glass did in 2013. They are accused of corrupting the morals of the young, a charge levelled at novels in the 1790s, motion pictures in 1910, comic books in the 1950s and video games in the 1990s. From 19th-century Luddites to modern prophets of robot-induced mass unemployment, the fear that new machines will deprive people of their jobs is centuries old. So too are concerns over new technologies that allow man to play god, from nuclear weapons to genetic modification to artificial intelligence; these are all modern-day versions of the myth of Prometheus, and the question of whether mankind could be trusted with the gift of fire. Whether such concerns are merited or not, an understanding of reactions to past technologies can give futurologists, entrepreneurs and inventors valuable clues about how new products might be received.

Tomorrow is another today

So much for history. The second place to look for glimpses of the future is the present. As William Gibson, a science-fiction writer, once memorably put it, "the future is already here – it's just not very evenly distributed". Technologies have surprisingly long gestation periods; they may seem to appear overnight, but they don't. As a result, if you look in the right places, you can see tomorrow's technologies today. This approach is taken by journalists and corporate anthropologists who want to understand new trends. It involves seeking out "edge cases": examples of technologies and behaviours that are adopted by particular groups, or in particular countries, before going on to become widespread. A classic example of an edge case is that of Japan and smartphones at the turn of the century.

In 2001, mobile handsets with cameras and colour screens were commonplace in Japan. They could display maps with walking directions and allowed users to download e-books, games and other apps. Journalists and analysts flocked to Japan to see these phones in action. And whenever Japanese visitors to European and US technology conferences passed around their handsets, they were treated as though they were artefacts from the future that had fallen through a rift in the space-time continuum. Japan arrived in the future early because of the isolated, proprietary nature of its telecoms industry; its domestic market was large enough to allow its technology companies to experiment with new ideas without worrying about compatibility with other countries' systems. It was several years before consumers in Europe and the US could buy handsets with comparable features. For a while *Wired* magazine had a column called "Japanese Schoolgirl Watch", predicated on the idea that what Japanese schoolgirls (the most ardent users of early smartphones) do today, the rest of us might be doing tomorrow.

Edge cases can arise in the most surprising places. Kenya, for example, has long led the world in the adoption of mobile money, which allows funds to be transferred from one handset to another instantly, as easily as sending a text message. For many years you could pay your taxi driver using your phone in Nairobi, but not in New York. Mobile money took off in Kenya in part because the lack of banking infrastructure offered a clean slate; in a country where most people do not have bank accounts, there is little competition from incumbent payment systems. Political factors also played a role: use of mobile money took off during the post-election violence of 2007–08, when it was seen as a safer alternative to banks, which were entangled in ethnic disputes.

And sometimes it is people who share a particular interest,

rather than those in a particular place, who pioneer the use of a new technology. The most obvious example is the technology community: geeks are the earliest adopters of new innovations, from e-mail to Uber. But they can also pioneer wider trends. Techie types led the way in adopting fitness-tracking devices, for example; the "quantified self" movement, which involves obsessively monitoring your health and fitness activity, started as a technological cult but has since gained a wider following.

Chris Dixon, a venture capitalist at Andreessen Horowitz, says he often looks to see whether a new technology or behaviour has spawned a thread on Reddit, an online discussion board. If it has, this suggests that it is gaining traction. For example, a growing number of techies are now interested in novel food technologies, from nutritionally complete food shakes (no need to cook, just chug) to caffeine-laced sweets that provide an edible alternative to coffee, though it is too early to say whether this sort of thing will catch on more widely.

Indeed, just as historical analogies are not perfect, looking at edge cases can also be risky. Some technologies never take off – or, when they do, they take off in an unexpected or different way. In the West, for example, smartphones initially followed the Japanese trajectory, but then took a completely different turn with the advent of the iPhone and other touchscreen devices. But what is undeniable is that all technologies that do eventually catch on first go through an underground period where their use is restricted to a subpopulation; they don't appear from nowhere. Finding these edge cases and identifying emerging technologies and behaviours is more art than science; trendspotting is hard. But it is the stock-in-trade of countless consultants and futurologists, not to mention technology journalists, who are always looking for new ideas and trends to write about.

The vision thing

The third place to catch glimpses of what is coming next is in the imagined futures of science fiction, whether in the form of books, television shows or films. Sci-fi stories take interesting ideas and carry them to their logical conclusions. What if we could build general-purpose robots, or a space elevator? What would happen if nanotechnology or biotechnology got out of control, or genetic self-modification became as commonplace as tattoos? Such futuristic tales provide visions of how the world might look with ubiquitous artificial intelligence, anti-ageing treatments that expand human lifespans, colonies on Mars and elsewhere in the solar system, or a fragmenting of humanity into post-human tribes. It can be a handy way to map out the space of potential long-term outcomes: what Elon Musk, a leading technology entrepreneur, calls the "branching probability streams" of the future.

Science fiction is not merely predictive, however. It also inspires technologists to invent things. Scratch a technologist and you'll find a sci-fi fan. The flip-open mobile phone of the 1990s, for example, seems to have been directly inspired by the portable communicators seen in *Star Trek* in the 1960s. More recently, the idea of being able to talk to computers, another idea from *Star Trek*, has inspired a new wave of computing devices, starting with the Amazon Echo, that use speech as their main interface, allowing always-on, hands-free use. Generations of computer scientists have grown up on Isaac Asimov's robot stories; today many entrepreneurs, including Musk, cite the Culture novels of Iain M. Banks as an inspiration. Like *Star Trek*, they depict a post-scarcity civilisation in which humans and artificial intelligences live and work together.

Yet although sci-fi is outwardly about the future in most cases, it is really about the present, and responds to contemporary ideas and concerns, such as an overdependence on machines or worries about environmental destruction. Reading a diverse selection of sci-fi can give you greater mental flexibility to envisage future scenarios, both technological and societal. But it can also unwittingly constrain, by shaping the way technological developments are perceived and discussed: robots, for example, look very different in the real world than they do in science fiction, and trying to imitate the fictional variety may steer roboticists in the wrong direction. So it is also worth reading classic sci-fi from the mid-20th century, to see what it gets wrong and why – and then ask yourself what mistaken assumptions are being made by today's tales.

Trying out the tools

We now have three tools – drawing on accounts of the past, present and future of technology – that can help us imagine the future. So let's put these tools into practice by briefly considering four worked examples. Each one is a technology that, at the time of writing, is emerging but still unproven; an edge case, in other words. What light can current trends, history and the imaged futures of science fiction cast on its likely trajectory? (Some of these examples will be treated in greater depth in subsequent chapters; the aim here is simply to show our predictive toolkit in action.)

Virtual reality

Having failed in the 1990s because the technology was not up to the job, virtual reality (VR) made a comeback in 2016. Several companies launched high-end headsets, driven by powerful PCs or games consoles, capable of transporting the wearer into an immersive, three-dimensional alternative reality. At the same time, a cheaper form of the technology emerged that involves using a smartphone in conjunction with a headset adapter.

What happens now? Current trends show a clear shift towards the smartphone, rather than the PC, becoming the most important device. So it seems likely that PC-based and console-based VR will be a transitional step, and the future of VR is as a smartphone-based medium. (Some people will be prepared to pay more for high-end VR systems, just as they do for high-end audio, but most people will not.) Today's headsets that work with smartphones are still clunky, like early mobile phones; but within a couple of years headsets could be much smaller devices that people routinely carry around, like sunglasses or headphones; so you could, for example, use a VR headset to watch a film, play a game or attend a virtual meeting while on a train.

As for the future, there is widespread agreement among techie types and sci-fi writers that augmented reality (AR), which mixes VR imagery with the real world, is likely to be the next big step in computer interfaces, beyond the touchscreen. A world in which imagery is overlaid on reality is a common sci-fi trope (often implemented using smart contact lenses or brain implants). But history offers a lesson: VR and AR are almost certain to cause a moral panic over their effects on children, just as films and video games did before them. Proponents of the technology could get ahead of this by commissioning research with which to address such speculation, and emphasising its educational and therapeutic uses, in addition to its potential in entertainment, communication and collaboration.

Autonomous vehicles

Cars capable of driving themselves – some of the time at least – are taking to the roads. There are two competing approaches: adding self-driving features to existing cars as driver aids, and building entirely new vehicles that can operate only in autonomous mode (fleets of which could operate as taxis in a city centre, for example, summoned using a ride-hailing app). Self-driving trucks are also under development.

There are many historical parallels with the advent of the motor car more than a century ago: concerns over safety, regulatory uncertainty, liability in the case of accidents and worries that the new technology will destroy jobs. The car displaced the existing travel infrastructure built around horses and carriages, and their related professions; but it also created new jobs for mechanics, drivers and workers in roadside service stations, restaurants and motels, and stimulated commerce more generally by making travel easier. A switch to autonomous vehicles, feared by truck drivers and taxi drivers, would have a similar dislocating effect, but also offers benefits in the long term.

Models suggest that shared, self-driving taxis could reduce the number of vehicles needed in a typical city by 90%. Most people would no longer need to own a car; space wasted on parking (as much as 20% of the area of some US cities) could be used for housing or parks; and autonomous cars could be electric, thus reducing climate-changing emissions. By reducing the cost of delivery, self-driving delivery vehicles might greatly expand demand for products (such as food) made locally. In developing countries, billions of people might skip car ownership altogether. Car accidents and deaths would plummet. Just as cars reconfigured cities in the 20th century, both history and current trends suggest that autonomous ones could do the same in the 21st.

Sci-fi has largely done a bad job of predicting all this, however. For narrative purposes, vehicles are more exciting if piloted by humans. But in future that seems likely to be the exception, not the rule.

Private spaceflight

In recent years the most progress in space technology has been made not by government space agencies but by private companies, notably SpaceX, a company founded by Elon Musk. It has pioneered the technology of reusable rockets, successfully landing the first stage of its Falcon 9 rockets on terrestrial landing pads and on ocean-going drone barges. This is significant because the first stage of a rocket accounts for around 70% of its cost, and is normally discarded in the sea after launch; recovering and reusing it could therefore dramatically reduce the cost of launches, and hence the cost of access to space. (Blue Origin, a rival company founded by Jeff Bezos, the boss of Amazon, has demonstrated the launch and reuse of smaller, suborbital rocket stages.) SpaceX's rockets are currently used to launch satellites into orbit and to deliver cargo to the International Space Station, but Musk has made no secret of his long-term goal: to establish a colony on Mars, as an insurance policy against a disaster wiping out humanity on Earth.

There is a rich tradition of science fiction about the colonisation of the solar system which examines in detail the complexity of establishing a human colony on Mars, and the political conflict that might then arise both within such a colony and between Earth and Mars. Discussion has already begun about what sort of political and legal systems might be introduced on Mars and on other colonies. But history can be informative too.

The obvious analogy is with the establishment of American colonies by British settlers, and their subsequent fight for independence. Other analogies are also worth exploring: during the era of Arctic and Antarctic exploration, for example, private expeditions were generally more successful and had higher survival rates than government-funded ones. The history of the gold rush may cast light on proposed asteroid-mining schemes. But perhaps the most striking analogy is with the development of aviation.

At the start of the 20th century heavier-than-air flight was considered impossible. Once it was shown to be feasible, it was considered dangerous. Then, starting in the 1930s, it began to evolve into an industry, initially catering for the rich. By the end of the century air travel had become widely affordable and was considered unremarkable. Given its current rate of progress, it is not unreasonable to assume that the space-travel industry might follow a similar trajectory, from crazy to commonplace, during the 21st century. Future generations may look back on the first two decades of the century as the era when spaceflight really began to get going, after the false start of the cold-war space race.

Human genetic modification

A family of genetic-editing techniques called CRISPR is causing much excitement among scientists and beyond. In essence, CRISPR is the genetic equivalent of "find and replace" in a word processor: it allows specific genetic sequences to be detected and edited with greater precision than previous techniques. This has enormous therapeutic potential. Embryos could, for example, be modified to remove genes that cause hereditary diseases, and people born from such embryos would not pass the diseases in question on to their offspring. But genetic therapy could easily tip over into genetic tweaking (for better eyesight, higher intelligence and so forth), raising the prospect of "designer babies". Debates about how best to regulate the technology are under way.

Science-fiction authors have already considered the long-term possibilities. If rejuvenation therapies allow people to live for hundreds of years, will only the rich be able to afford them? Should people be allowed to modify their bodies to give themselves wings, or gills and fins? Rather than modifying other planets to support human life, might it make more sense to modify humans to allow them to live in different environments? The idea that mankind might split into several distinct post-human species is a common sci-fi trope. Some people might prefer to transplant their brains into robot bodies; others might prefer to modify themselves to assume non-human forms.

In the short term, expect debates about access to genetic therapies which may echo historical arguments about expanding access to vaccines and HIV/AIDS treatments. Arguments over genetic self-modification, and the extent to which people have autonomy over their own bodies, may come to be seen as extensions of current debates about the right to doctor-assisted dying. Human rights have been extended in many respects over the past century: this is likely to be part of the battleground in the next century.

A faster future?

These are just four areas where current trends, historical examples and science fiction suggest there is great potential for progress, and associated upheaval, in the next few decades. Taken together, they suggest a broader analogy with the scientific revolution of the mid-17th century. It was a time when new tools and technologies, notably the microscope and the telescope, came together with new scientific and mathematical methods. Natural philosophers (the term "scientist" was a 19th-century coining) realised the extent of their ignorance in a range of fields, from physics to biology, resulting in a fertile period of discovery and invention.

In many ways the current state of science and technology feels similar: it is clear that contemporary understanding of the principles of genetics or artificial intelligence is rudimentary, for example, and decades of work must be done to unravel them. Just as 17th-century advances in mathematical theory helped scientists in a wide range of fields, the same may be true of today's information-processing techniques, such as "big data" tools and machine-learning systems.

There is also huge potential for cross-fertilisation between previously disconnected fields. Genetics has made biology and medicine into information sciences, for example. And there is a growing volume of two-way traffic across the border between the fields of neuroscience and the structure of the brain on the one hand, and computer science and the construction of artificial neural networks on the other.

The scope for rapid progress is unprecedented in some ways, but feels familiar in others. We have new frontiers to explore, and new tools with which to explore them. In their imagined futures of the 21st century, two science-fiction writers, Kim Stanley Robinson and Charles Stross, have even given this emerging scientific revolution a name: the Accelerando.

Of course, predicting exactly how the future will play out is impossible. But, if you look in the right places, it is possible to make some educated guesses along the way.

2

Physical foundations of future technology

Frank Wilczek

Advances in fundamental physics have created a qualitatively new situation for its relationship with technology. Secure foundations allow us to perceive both limitations and opportunities clearly. Brilliant prospects lie ahead, but also dangers

FUNDAMENTAL PHYSICS both constrains and enables technology. Abstractly this is a truism, since much of technology is embodied in machines and structures which, being physical objects, are subject to the laws of physics. Yet over much of history, in almost all areas of technology, the relationship between fundamental theory and practical applications has been rather loose. Consider, for example, some outstanding highlights of Roman engineering: their great roads, aqueducts and the Colosseum. As described by Vitruvius, in *De Architectura*, the technology that supported those feats was based on long-accumulated experience, codified in empirical rules of thumb. We find, for example, detailed instructions on the choice of construction materials and their preparation – in some ways, anticipating the composites of today – but there is nothing that we would recognise as systematic materials science. Similarly, the central motif of Roman construction, the arch, is presented as a template, not as a mathematically determined solution to problems of loading and stress. (And the template, based on circular segments, is not optimal.)

Today the connection between fundamental physics and technology is much tighter. Notably, modern microelectronics and telecommunications support the processing and transmission of information at speeds that would have seemed utterly fantastic just a few decades ago. Those profoundly enabling technologies would be unthinkable without deep, reliable understanding of the quantum theory of matter and of light (including radio, microwaves and the rest of the electromagnetic spectrum). No amount of tinkering or "innovation" could get you there.

My main goal, in this brief essay, is to survey the current state of fundamental physics in so far as it is relevant to technologies that might emerge over the next 50 years. This will also suggest future directions and opportunities.

From mystery to mastery

Let me begin with a central claim, which I will then both explain and defend:

> *We have, today, accurate, complete equations adequate to provide the foundation of nuclear physics, materials science, chemistry and all plausible forms of engineering.*

Thus in principle we could, by solving the appropriate equations, replace experimentation with calculation, in all those applications. This represents, in human history, a qualitatively new situation. It has arisen over the course of the 20th century, primarily as a result of dramatic advances in the application of quantum mechanics.

To gain perspective on the central claim, it is illuminating to consider a bit of history.

At the beginning of the 20th century there were many basic, supremely important features of the natural world that fundamental physics could not account for. Chemists had deduced, empirically, a periodic table of elements. They had also built up a richly detailed picture of the geometry of molecules, notably including the ring structure of benzene and other organics, and used it successfully to design new molecules and reactions. But the known laws of physics did not explain the existence of stable atoms, much less their properties, and still less the formation of chemical bonds. Similarly, the basic properties of materials, such as their electrical conductivity, strength and colour, could not be connected to fundamental physical laws.

The sun's source of energy was utterly mysterious, and its rate of cooling, as calculated by Lord Kelvin, was too abrupt to be consistent with Charles Darwin's evolutionary biology. It was an open question whether basic phenomena of life (metabolism and reproduction) and thought (cognition) could possibly emerge as consequences of the normal behaviour of physical matter, or would require additional "vital" ingredients.

Over the course of a few decades all these problems were solved convincingly. They yielded, not to direct assault, but to a determined application of the strategy Isaac Newton called "analysis and synthesis", a method now often given the semi-pejorative label "reductionism". According to this strategy, we progress by attaining rigorous understanding of the properties and interactions of basic parts (analysis) and then leveraging that understanding to deduce the properties of more complex assemblies mathematically (synthesis).

In retrospect, we can identify two events near the dawn of the 20th century as pivotal. One was the discovery by J.J. Thomson, in 1897, of electrons as a widely present component of matter. Electrons have distinctive properties, the same for all electrons anywhere, observed at any time. In those respects, they are the archetypal "elementary particle". Because electrons turn out to obey ideally simple equations, they are still regarded as elementary particles today. Electrons are the main players in chemistry, and of course in electronics.

The other was the introduction by Max Planck, in 1900, of an irreducible unit, or quantum, of action: Planck's constant h (technically, action = energy × time). Planck introduced his constant in the course of a recondite discussion of the thermodynamics of radiation, and his use of the constant was limited to that context. It was Albert Einstein, in 1905, who interpreted Planck's constant to mean that light travels in irreducible particle-like units, which today we call photons. The photon is a second elementary particle. An important philosophical implication of the Planck–Einstein ideas is to ease the distinction between light (built of particles) and other kinds of matter (built of particles). That implication has stood the test of time. In the remainder of this article, when I say "matter" I will use the term inclusively, to include light.

The next great step was to achieve a physically grounded model of atoms. It occurred in the years 1911–13, and involved both experimental and theoretical components. In 1911 Hans Geiger and Ernest Marsden, at the suggestion of Ernest Rutherford, studied the deflection of fast-moving alpha particles by gold foil. The unexpected ability of the gold to produce large deflections, as analysed by Rutherford, indicated that all of the positive electric charge and almost all the mass of a gold atom is concentrated in a tiny core, or nucleus, which occupies about a millionth of a billionth of the atom's volume. One might imagine, with Rutherford, that electric forces would bind electrons to the nucleus, forming the atom. But that plausible picture could not be reconciled with the known laws of physics. No effect was known that could prevent electrons from spiralling into the nucleus. Here Niels Bohr, in 1913, proposed that only a very restricted class of orbits were possible, in blatant contradiction to Newtonian principles. Bohr's criterion for which orbits are allowed introduced Planck's constant, until then a standard applied to photons, into the description of electrons.

The Bohr model was so boldly simple and, in its application to hydrogen, so strikingly successful that Einstein called it "the highest form of musicality in the sphere of thought". It was not, however, formulated in equations that could be applied to other problems. And since its hypotheses contradicted the principles of macroscopic (Newtonian) mechanics, it was not at all clear how to embed Bohr's ideas into mathematically consistent, widely applicable equations.

Over a decade of struggle several physicists made important contributions to solving that problem, but in this narrative I will (over-) simplify to say that in 1925 Werner Heisenberg got consistent equations for electrons described as particles, while in 1926 Erwin Schrödinger got consistent equations for electrons described as waves. At first the relationship between Heisenberg's equations and Schrödinger's was not apparent, but Paul Dirac, also in 1926, showed that they were, in their consequences, mathematically equivalent, and could both be derived from a common, more general starting-point. Dirac's mathematics could accommodate both electrons and photons. His theory of the interaction of electrons and light, quantum electrodynamics (QED), successfully covered such a wide range of phenomena that he would already claim, in 1929, that:

> *The underlying physical laws necessary for the mathematical theory of*
> *a large part of physics and the whole of chemistry are thus completely*
> *known, and the difficulty is only that the exact application of these*
> *laws leads to equations much too complicated to be soluble.*

This statement is, of course, ancestral to our central claim.

In the 1940s experiments in atomic physics achieved such high precision that new, more accurate methods of solving the basic equations of QED were required, in order to test that theory rigorously. These new methods, developed by Julian Schwinger, Richard Feynman, Sin-Itiro Tomonaga and Freeman Dyson, showed that QED describes the behaviour of electrons in a wide range of conditions (including all those relevant to chemistry and engineering) with accuracy better than a few parts per billion.

After these triumphs the outer parts of atoms were adequately understood, but their nuclei remained mysterious. The deep theory of subnuclear forces – the so-called strong and weak forces – emerged in the 1970s, and was tested rigorously in the 1990s, completing the "effective theory" of matter we use today. But now let us pass from narrative to description.

System of the world

Non-physicists are often bemused to hear physicists speak of the "simplicity" of their fundamental theories. For in practice only a very small proportion of the human race understands those theories, and it takes years of determined study and hard thinking for any individual human being to achieve understanding. Yet there is a precise, profound sense in which the theories are simple, and it is important to appreciate this feature. It forms an essential supplement to our central claim, defining its use of "complete".

The equations of fundamental physics can be encoded in a short computer program. By following the instructions of the program a computer, with no further input from the external world, will be able (given sufficient time) to calculate their consequences unambiguously.

As far as I know no one has actually written such a program. It would be an interesting exercise. I estimate that in a high-level computer language, such as *Mathematica*, no more than a few hundred lines of code would be required. (Please note that efficient encoding, which enables rapid solution of the equations in interesting applications, is a different, and probably open-ended, problem.)

Basic principles

The fundamental equations of physics, as presently understood, are codified in four related core theories, governing four fundamental forces: gravity, electromagnetism, and the strong and weak forces. Their combination is often called the standard model. They embody, and can be derived from, three basic principles: relativity, gauge invariance (also called local symmetry) and quantum mechanics.

Two of these principles, relativity and gauge invariance, are statements of symmetry. The word "symmetry", in this context, is used to mean "transformation without change", or, more elegantly, "change without change". A circle illustrates the basic concept. We can transform a circle, by rotating it around its centre. Each point on the circle moves, so this is a genuine transformation, but the circle as a whole does not change. Similarly, the central assumption of special relativity is that one can transform the properties of all the objects in the physical world, by moving at a constant velocity (thus altering the apparent velocity of all the objects we view), without changing the laws those objects obey. Gauge invariance involves other transformations, which bring in much less familiar properties than velocity, but invokes the same idea. We constrain the laws, by demanding that they operate in the same way in a wide variety of situations.

The third principle, quantum mechanics, is a broad framework rather than a specific hypothesis. In this aspect it resembles classical (Newtonian) mechanics, which tells you how motion results from given forces, but does not tell you what forces there are. In detail, the ambiguities of quantum theory are more severe. (For experts: here I allude to alternative choices of dynamic variables and to so-called ordering ambiguities.) Thus the application of quantum mechanics to specific physical problems, prior to the emergence of our core theories, always involved some guesswork. But the core theories enforce unique choices, for deep reasons I will indicate in the next paragraph. Although it is not widely appreciated, I think it is fair to say that it is only in the context of our core theories that we come to understand exactly what quantum mechanics is.

It turns out to be quite difficult to formulate equations that are consistent both with relativity and with the principles of quantum mechanics. The theories that do this, relativistic quantum field theories, contain many basic quantities that are ill-defined or, formally, infinite. Only carefully crafted combinations, for which the infinities cancel, can be used in modelling the physical world. In order to get enough of these, we must implement the framework of quantum mechanics in a very specific way. All ambiguity disappears. The fact that the theories which implement our basic principles are poised on the edge of inconsistency makes them hard to find. But it offers, in return, the gift of rigidity. It leads us to very specific equations and procedures whose resistance to change ensures their staying power.

Emergent principles

Two consequences of the fundamental laws are so basic and all-important that they deserve mention, even in this brief survey.

The primary objects in nature are space-filling, persistent (ie, time-filling) fields. Particles, such as electrons, are excitations of the corresponding field. Thus all electrons have the same properties, wherever and whenever they are encountered, because each is an excitation of the same field. The exact sameness of all electrons (and of other elementary particles) is profoundly significant. The development of interchangeable parts was a great step forward in 19th-century industrial technology, allowing mass production, assembly and repair. Chemistry, biology and engineering as we know it all rely on nature's abundant supply of interchangeable parts.

When electrons and atomic nuclei come together to form an atom, or when quarks and gluons come together to form a proton, the resulting object has a unique, stable structure, which cannot be distorted unless a significant amount of energy is deposited. (This contrasts with systems based on classical mechanics, such as solar systems, which can absorb arbitrarily small amounts of energy, at the cost of small changes to their structure.) This "quantum censorship" means that we can, in appropriate circumstances (when there is not much energy in play), treat an atom or a proton as a black box, whose internal structure is hidden. Thus, for example, we need not worry about quarks and gluons when we're designing a transistor.

Those two emergent features of the fundamental laws allow us to build up our synthesis of nature stepwise, and to use statistical considerations when dealing with large numbers of (indistinguishable) entities. In this way, they put many common practices of chemists and engineers on a sound footing, as emergent consequences of a deeper "reduction".

Building blocks

Standard elementary descriptions of matter present protons and neutrons as the building blocks of atomic nuclei. Electrons then fill out the bulk of atoms, and atoms combine into molecules and materials. To reflect the state of the art, this elementary description needs a few refinements.

First, as mentioned earlier, we now realise that it is unnatural and unnecessary to separate light from matter. Photons should be added in.

Second, we must get past the idea that protons and neutrons form appropriate ingredients for fundamental work. Experiments reveal that protons and neutrons are complex objects, with elaborate internal structures. The more basic particles, from which protons and neutrons are made, are called quarks and gluons. All available evidence confirms that quarks and gluons obey the ideally simple equations of quantum chromodynamics (QCD), as described above. There are two important kinds of quarks, called up (or u) and down (or d).

Third, we must include (electron) neutrinos. These particles are emitted in the processes of nuclear transformation that power the sun, and are exploited in various nuclear technologies (including medical diagnostics, some forms of radiation therapy, nuclear power reactors and nuclear weapons).

With electrons, photons, gluons, u and d quarks, and (electron) neutrinos as fundamental ingredients, we have enough to construct an "effective theory" that satisfies our central claim. It is constructed from a much more economical list of ingredients than the traditional periodic table, comes equipped with a much more precise instruction manual (its fundamental equations) and covers a much wider range of phenomena.

Our effective theory has known limitations, as I shall discuss presently, but they do not appear relevant, for the foreseeable future, to any plausible, broadly significant technology.

Cosmic resources

The past few decades have been a golden age for physical cosmology. Evidence for a surprisingly simple history of the universe, starting from a nearly homogeneous hot big bang and acquiring structure through gravitational instability, has become both precise and overwhelming. It would be out of place to recount that story here, but two consequences are especially relevant to our central themes, and deserve explicit mention.

Our effective theory tells us about the different forms that matter can take, but it does not, in itself, tell us what materials are actually available. The big-bang hypothesis, according to which the universe began very hot, implies that nuclei congealed from a primordial mix of quarks and gluons, and allows us to calculate the relative abundances of different chemical elements in the early universe, prior to star formation. The result is, overwhelmingly, hydrogen and helium. The heavier elements are cooked inside stars and then liberated as stars die, in supernova explosions. Following out this scenario, we obtain an excellent account of the material we find in the universe today. This agreement between fundamental physical theory and observation further reinforces our confidence in the theory, even in applications to conditions that are far more extreme than occur in terrestrial chemistry or biology, or are contemplated in engineering.

However, astronomers have also gathered convincing evidence that ordinary matter, based on electrons, photons, gluons and quarks, constitutes only about 4% of the universe by weight. The rest is categorised as "dark matter" (about 25%) and "dark energy" (about 70%). Both dark matter and dark energy have been identified, so far, only through their feeble (but cumulative) gravitational influence on ordinary matter. Because the interactions of dark energy and dark matter with ordinary matter are extremely feeble – so feeble as to have escaped heroic efforts at detection – it is difficult to imagine how they could become useful resources for technology.

Loose ends and omissions

Complications: more stuff

The most determined pursuit of analysis and synthesis, or reductionism, is carried on at great accelerators, such as the Large Hadron Collider (LHC) at CERN, Europe's particle-physics laboratory. At the LHC, protons are accelerated to enormous energy, and then allowed to collide. Such collisions produce, fleetingly, energy densities far beyond any that occur naturally on Earth (or, as far as we know, anywhere else in the present-day universe). They allow us to test, with quantitative rigour, our theories of fundamental interactions under far more stressful conditions than they will encounter in practical applications.

For our purposes, the most notable result of this work has two aspects. For the effective theory, there is good news and bad news – but the bad news is superficial, whereas the good news is profound.

First the bad news: our "effective theory" is seriously incomplete. To get a good description of all the phenomena discovered at accelerators, we need to add four more kinds of quarks (strange s, charm c, bottom b, top t); two heavier electron-like particles (muon μ, tau lepton τ), each of which introduces, in addition, its own neutrino; two heavy relatives of the photon and gluons (W and Z bosons); and finally the recently discovered Higgs particle.

The bad news is that reality, viewed up close, contains unexpected complications.

Now the good news: the complications serve to reinforce the principles of the effective theory, and they do not compromise its practical applications. Investigation of the new particles provides many opportunities to test the general principles that underlie our effective theory – relativity, quantum theory and local symmetry – in demanding new ways. Indeed, these principles predict the rate at which the particles will be produced in different circumstances, the sorts of things they will decay into and many other details. So far these predictions, without exception, have proved to describe reality correctly.

We can thereby infer, with some confidence, another prediction: that the effects of these particles in normal (non-accelerator) terrestrial environments are negligible.

The good news is that the added ingredients, beyond our effective theory, are well understood, quantitatively. Their observed behaviour reinforces the validity of our general principles. But they are difficult to produce, and for the most part (with the exception of the new neutrinos) highly unstable. Their practical impact has been, and almost certainly will continue to be, negligible.

Quantum doubts, quantum gravity

Many of the pioneers of quantum theory, notably including Planck himself, Einstein and Schrödinger, were unhappy with its mature form. They were uncomfortable with its inherently probabilistic predictions, and with its insistence that in the subatomic world "perfect" measurements – that is, measurements which do not affect the system being measured – are not merely an idealisation, but a physical impossibility. These features of quantum theory seem to undermine the notion that there is an objective world, containing objects with definite properties, which evolves according to definite principles.

Later generations of physicists, for the most part, have made their peace with quantum theory. It has powered many new advances and survived every new test. In addition, technical work around the notion of "decoherence" has clarified how the stable, essentially deterministic behaviour of large bodies can emerge from quantum behaviour in the micro-world. But even today some highly accomplished physicists perceive difficulties in the foundations of quantum theory. (I don't.) Designs for quantum computers make heavy use of the strangest and most delicate features of quantum theory. It would be very interesting if they unexpectedly fail.

Since the difficulty, in particular, of fully reconciling our theory of gravity, general relativity, with the principles of quantum mechanics has been the subject of much hyperbole, it is important to bring that discussion down to earth. At a practical level, there is no problem. Astrophysicists and cosmologists routinely, and successfully, calculate behaviour in physical situations where both gravitation and quantum theory are in play simultaneously. Throughout that work, no significant ambiguities or singularities arise.

Problems do arise if we try to apply the equations to such extreme conditions as might occur during the earliest moments of the big bang, or in the deep interior of black holes, where their solutions become singular. Conceptual puzzles also arise in the quantum theory of small black holes.

But it would be a bracing achievement, and major progress, to identify any concrete, observable phenomenon which brings in truly characteristic features of quantum gravity, and of course to observe it. So far, despite widespread, intense focus and the prospect of fame and fortune, no one has risen to that challenge successfully.

Things that don't happen

An important function of fundamental understanding is to spare us from lines of thought that are likely to be unproductive. Here I would like to mention three potential "technologies" that are amply represented in popular media but are in considerable tension with fundamental physics. Of course surprises are always possible, and nature is the ultimate authority, but developments along the following lines would require us to unlearn principles that, so far, have served us very well:

- Faster-than-light information transfer violates special relativity. In extreme conditions, where there are strong gravitational fields, space-time can be warped, and there might be shortcuts (wormholes) that connect otherwise distant points. But as I shall discuss below, usable wormholes seem far beyond the means of any achievable technology.

- Long-range influences, as envisaged for example in astrology, are not part of our standard model. Not coincidentally, they also run counter to the accumulated experience of science, that even exquisitely delicate experiments yield reproducible results, independent of the external world, once we take a few routine precautions.

■ Mental powers, explicitly envisaged in extrasensory perception, telekinesis, clairvoyance and so forth, or implicitly invoked as "consciousness", divorced from a physical substrate, are subject to similar comments. They have no place in today's fundamental physics; and even in the most exquisitely delicate measurements, experimenters have not found it necessary to concern themselves with what people are thinking.

Opportunities

Computing reality

We can look forward to the day, perhaps not long in the future, when computers do for nuclear physics, stellar physics, materials science and chemistry what they have already done for aircraft design, supplementing and ultimately supplanting laboratory experimentation with computation.

The recent development of QCD, our theory of the strong interaction, gives a foretaste of things to come. The initial validation of the theory came through its accurate quantitative description of processes at very high energy, where the behaviour of the theory simplifies. But nuclear physics, which motivated people to study the strong interaction in the first place, is a tougher nut to crack. Much ingenuity was directed towards solving the equations of QCD by analytical techniques. But the most successful approach, by far, has been to put the fundamental equations in a format that computers can run with, and then to let them run. Now we can anticipate a future when nuclear physics reaches the level of precision and versatility that atomic physics has achieved today. A refined nuclear "chemistry" could give us ultra-dense stores of energy that are smaller, better controlled and more versatile than today's reactors (or bombs).

Calculation will increasingly replace experimentation in the design of useful catalysts and drugs, leading to much greater efficiency, and opening up new opportunities for creative exploration.

Many questions of technology hinge on properties of materials. More efficient batteries (energy supplies) could revolutionise robotics; more efficient photovoltaics could ease the transition to large-scale use of solar energy; room-temperature superconductors could enable frictionless rail transport; strong materials could allow us to build a space elevator, linking Earth to space cheaply and reliably. In each of those important applications, and many more, relatively small improvements in key material properties could change the ground rules dramatically. Can it be done? Our equations contain the answers – but to extract them we must compute.

There are two issues here: hardware and software. Twenty-five cycles of Moore's law (see Chapter 4) have given humans in general, and physicists in particular, computational tools of extraordinary capacity. The pace of exponential expansion is levelling off, no longer involving a doubling of the number of components in an integrated circuit every two years or so, as miniaturisation reaches atomic scales and new physical rules apply. Still, we can anticipate at least a few more cycles in coming decades, even without a drastic turn from existing semiconductor technology.

There are also several promising new directions in view. Most present-day information processing is based, fundamentally, on moving electric charge (embodied in parcels of electrons) around. But electrons move much slower than light, and their motion generates heat, which is troublesome to remove. Light is already used routinely to transmit dense information over long distances – a process that requires conversion from electronic coding to light coding and back. The converters are rapidly becoming more efficient and versatile in their operations on light, and could evolve into free-standing "photonic" computers.

More revolutionary is the prospect of quantum computers, which encode information in the form of delicate correlations (entanglement) among quantum systems. In principle these correlations have a very rich structure, so that extraordinary densities of information can be stored and manipulated. Unfortunately, it is difficult to overstate how delicate complex entanglement is. Several possible technologies to protect and exploit it appear promising, but they are in their infancy. If large, practical quantum computers can be built, they should be very good at solving problems in quantum mechanics, unleashing the latent power of our central claim.

Yet another direction takes inspiration from biology. Present-day mainstream computers are essentially two-dimensional. They are based on chips that must be produced under exacting clean-room conditions, since any fault can be fatal to their operation. And if they are damaged, they do not recover. Human brains differ in all these respects: they are three-dimensional, they are produced in messy, loosely controlled conditions, and they can work around faults or injuries. There are strong incentives to achieve these features in systems that retain the density, speed and scalability of semiconductor technology, and there is no clear physical barrier to doing so.

Efficient algorithms exploit special features of the problems they target. Their advancement is an inherently creative process, and difficult to address in generalities. Here I will mention just one aspect of software development, which is especially remarkable. Much of the work that went into sustaining Moore's law, especially in its more recent cycles, was informed by sophisticated software and CAD (computer-aided design) tools which, by unfolding the implications of fundamental physics in new circumstances, allow engineers to explore and refine new ways of implementing circuit elements (for example, miniaturising transistors) and to optimise circuit architecture. Thus there has been a powerful feedback loop, whereby advances in computation lead to better computer designs, which lead to advances in computation. With the increasing sophistication of artificial intelligence, we can anticipate many other feedback loops of this kind, whereby insights from more capable (and increasingly autonomous) computers lead to creation of yet more capable computers.

Expanding reality

Fundamental physics informs us that there are important aspects of the world we might, but do not yet, observe. I will mention a few outstanding possibilities.

Recently an extraordinary instrument, the Laser Interferometer Gravitational-Wave Observatory, or LIGO, which is operated by Caltech and MIT, reported observation of a signal due to gravity waves associated with the merger of two massive black holes, each weighing a few tens of solar masses. LIGO is designed to detect extremely tiny changes in the distances between a few pairs of mirrors. The numbers are mind-boggling. The mirrors are 4 kilometres apart, and the distances between them are expected to change by less than one-thousandth of the diameter of a proton. All kinds of things can jiggle mirrors, but gravitational waves produce a unique pattern of changes, so their signal can emerge from the noise. This observation was the culmination of a 50-year effort. It would, of course, have been unthinkable without guidance from fundamental physics as to what sort of signals to expect, and how to measure such tiny distances. (*Inter alia*, the fact that such catastrophic events produce such small distortions in space-time dampens dreams of engineering wormholes, warp drives, time machines and the like.)

Gravitational waves open a new window on the universe, allowing access to hidden regions and violent events. To exploit its full potential, we shall need to deploy arrays of precision instruments, spanning millions of kilometres, in space.

Closer to home, human perception leaves a lot on the table. Consider, for example, colour vision. Whereas the electromagnetic signals arriving at our eyes contain an infinite, continuous range of frequencies, and also polarisation, what we perceive as "colour" is a crude hash encoding of a single octave, where the power spectrum is lumped into three bins, and polarisation is ignored. Many animals do finer sampling, and extend their sensitivity into the infrared and ultraviolet. We humans perform a much more refined frequency analysis of sound, and can resolve many distinct tones within chords.

There is valuable information about our natural environment, not to mention possibilities for data visualisation and art, on offer here. Modern microelectronics and computing offer attractive possibilities for accessing this information. By appropriate transformations, we can encode it in our existing channels, in a sort of induced synaesthesia. We will vastly expand the human sensorium, opening the doors of perception.

With more powerful sensors and actuators, out-of-body experiences will become more compelling. It is easy to imagine brilliantly attractive possibilities: immersive tourism to anywhere, any time, without needing to leave home. Fragile human bodies are ill-suited to deep-space environments, but human minds will experience them richly. Astronomy will be advanced by an expanding swarm of robotic probes, virtual telepresence and appropriate biological seeds much more readily than by sending fragile, ill-adapted human bodies into deep space.

Biology as resource and inspiration

If we take our central claim at face value, then biology supplies "existence proofs" for potentialities of matter that would otherwise be far from obvious. I have already mentioned the feasibility of sophisticated three-dimensional, self-assembled, self-repairing information processors. This might have seemed a vague fantasy, but for the fact that most human skulls contain fully realised examples. Similarly, the possibility of extremely parallel, rapid processing of massive data streams based on slow, unreliable circuit elements might seem a visionary dream, but for the actual performance of the human visual system.

Biology, of course, inspired John von Neumann's pioneering designs for self-reproducing "universal constructor" machines. Unlike his designs for computer architecture, which have become the foundation of a world-changing technology, self-reproducing (and – why not? – evolving) machines remain, at present, an intellectual curiosity. But the human race itself provides an existence proof for its potential. Our vastly improved understanding of how nature does it, at the molecular level, and our vastly improved ability to control flows of matter and information, specifically in 3D printing (see Chapter 10), could (re-) vitalise this awesome concept.

Conversely, nothing in the nature of matter itself suggests that ageing or disease is intrinsic to humans as physical beings. Our understanding of, and ability to monitor and control, matter at a fundamental level should enable us to overcome these infirmities. In practice, they pose a multitude of challenging problems and will continue to inspire new ventures in microscopy, broadly conceived, and in data analysis, enabling sophisticated diagnosis, and in molecular engineering, enabling sophisticated treatment.

Summary: failure modes

Our mature, profound understanding of matter suggests brilliant opportunities for creativity in the service of human ends. For reasons I have sketched above, as a physicist I feel confident in asserting that our newly profound understanding of how the world works opens up inspiring prospects for achieving new levels of material wealth and spiritual enrichment. We know what is possible, and we deduce that there is a lot that has not yet been done.

Before closing, however, a cautionary note is in order.

Since modern technology ensures stable documentation and widespread dissemination of knowledge, both existing and newly created, one might be tempted to think that technological history, and therefore ultimately human history, is now safe against significant backsliding. Progress may be uneven, but regress is off the table.

Or is it? Three failure modes arising from modern technology itself seem to me especially worrisome: nuclear warfare, ecological collapse and artificial-intelligence (AI) warfare.

Familiarity, and 70 years of good fortune, should not numb us to the horrific potential of nuclear weapons. They continue to exist in large numbers – many thousands – controlled by nine separate nations. Watch *Threads*, followed by *Dr Strangelove*. Enough said.

Ecological collapse due to anthropogenic climate change is another catastrophic risk. Atmospheric carbon pollution accumulates slowly on political timescales, and an appropriate response (ie, accounting for externalities) would require discounting many trillions of dollars in perceived assets. The determined resistance of asset holders may be difficult to overcome. It is an open question whether humanity can muster the maturity and wisdom to address this insidious problem.

Neither human nor artificial intelligence can evade David Hume's insight that no moral "ought" statement can emerge from an accumulation of logical or scientific "is" statements. He famously concluded: "Reason is, and ought only to be the slave of the passions, and can never pretend to any other office than to serve and obey them."

The creators of capable autonomous agents (such as humanoid robots) will, whether explicitly, by programming, or implicitly, by design choices, set the autonomous agents' underlying goals and motivations, their "passions". Many proposed applications of AI are meant to serve humanity in straightforward, benign ways. They can best be achieved through autonomous agents whose goals and motivations are likewise straightforward and benign. Much more problematic, however, is the use of advanced AI for military purposes: think of robot armies or, more generally, highly capable weapons systems like Strangelove's Doomsday Machine, set to perceive and act upon threats without human intervention. One can anticipate that super-intelligent entities designed to be suspicious and aggressive will manifest their suspicion and aggression in highly creative, unexpected ways. Competing groups of such entities could engage in uniquely effective warfare, leaving humans and their civilisation as collateral damage.

Note

This work is supported by the US Department of Energy under grant Contract Number DE-SC0012567.

3

Biotechnology's possibilities

Robert Carlson

**From medicine to manufacturing, biology will open up
extraordinary possibilities for individuals, companies and
economies**

IN 2050 *THE ECONOMIST* may be beamed directly into your
brain. The Defence Advanced Research Project Agency (DARPA), the
research arm of the US Defence Department, wants to build a bridge
between digital devices and the human cerebral cortex. The $60
million project has an audacious goal: to create universal digital input
and output functions for the brain. It is hard to anticipate what might
ultimately result from this integration, but it is clear that the future
of biology is not limited even to what we imagine as possible today.
The explicit, direct fusion of electronic and biological computation
will expand in new and unpredictable directions the capacities of both
animate and inanimate matter.

DARPA programme managers sometimes call this neural interface a
"cortical modem". The project builds on substantial clinical experience
gained from wiring up human neurons to electronic devices. Cochlear
implants and artificial retinas have been in use for years, restoring
hearing and sight respectively. Clinical trials of electrode arrays
implanted in brains have been used to reroute neural signals around
spinal injuries and to enable the direct neural control of robotic limbs.
In some cases, these neural prostheses have enabled paraplegics to
walk again.

Yet even as researchers make real progress in building the bionic
woman, this disruptive technology is itself being disrupted. Cybernetic
repair of biological functions will soon be competing with tissue

regrowth and replacement. Tissue engineers are building or growing transplantable internal organs, bones and connective tissue. Some of these innovations have yet to graduate from the research lab. Others are already in the clinic, including such varied examples as bladders, hip joints, vaginas, windpipes, veins, arteries, ovaries, ears, skin, the meniscus of the knee and patches for damaged hearts.

Progress is accelerating through the development of increasingly diverse methods to construct tissue. Just as the cortical modem would expand biological potential to include direct access to digital computation, the fruits of digital computation are enabling new means of manipulating cells. Would-be organ engineers, once limited to coaxing tissue growth from individual cells in a dish, can now use 3D printers to position cells accurately on organ-shaped scaffolds, like parts in the chassis of a car, aircraft or smartphone. Viewed as another sort of chassis, our bodies will soon become maintainable by swapping out diseased or exhausted tissues for new ones. Our wetware will become upgradable.

Constructing replacement tissues is just the first step in developing regenerative medicine as a discipline that will initially improve the quality of life, and then probably extend it significantly. Therapies that address the molecular mechanisms of ageing will gradually emerge over the coming decades in small steps that, viewed individually, may appear incremental. There may never be a day when we say "we are finished", but the impacts will accrue, and compound, over time.

Beyond anatomical and pharmaceutical interventions, humans will soon begin editing their own genomes. We will first seek to eliminate from human experience certain diseases, starting with those caused by genetic variations that are relatively easy to identify, such as beta thalassemia, Huntington's and sickle-cell anaemia. We will follow on by reducing the risks of Alzheimer's disease, cancers of all types and heart disease.

Today many observers fret over genetic modifications made to non-viable zygotes for research purposes, but the implications are more immediate than whether we choose to modify, or not, the genomes of future generations. Individuals will soon have the option to alter their own genomes for whatever purpose they desire. It requires no stretch of the imagination to list many ways that humans might wish to genetically improve their basic health, mental and physical performance, and appearance. We will be occupied for decades with ethical conversations around how far to take such modifications, and about who has a right to have access to them or the right to deny them to others.

Regardless of the course of this debate, demand is likely to drive adoption faster than policymakers can address the issues. That demand will be rooted in the desire to augment inherited human mental and physical potential with technology. We see this desire manifested every day in cosmetic surgery, tattoos, vision correction, and the use of performance-enhancing drugs in sports and in the classroom. Thus, even as the repair of bodies may be biological, the demand for expanded capabilities is likely to drive the development and deployment of technology that includes the cortical modem.

The connected self

A neural interface would provide direct connections between the human mind and the internet. Through these links, all available physical, electronic and economic infrastructure would become a suite of prosthetics that lends enormous power to connected individuals. We would extend our reach and touch across the globe with networked robots. We would expand our thoughts via direct access to libraries, supercomputers and space telescopes. Rather than peering through narrow chinks in our craniums, our minds could be truly exposed to the universe, with all the benefit and risk that might entail.

The cortical modem would be, by definition, a two-way communications channel. Connecting this interface to human neurons raises the question of what else might be embedded in the ensuing information flow. If we start streaming the internet into our brains, we will also be importing every network-security problem we face today, and presumably some that have not yet been invented. Spam and malware on our smartphones will be the least of our concerns. DARPA is already aware of the potential problems, having recently limited the strength of a fully articulated bionic forearm replacement to "human normal", specifically because of security concerns for the networked device. The agency was concerned about hacking the command routines of a super-strength arm, whether by its wearer or by a foe. This is just the beginning.

Who will hold the password to your cortical modem? Who will control the installation of the inevitable stream of software updates? As security agencies today insist on access to phone calls, e-mails and the contents of your phone or laptop, in 35 years will the government insist on a backdoor to your brain? How often will other parties exploit those same backdoors? Even in the face of these questions and risks, many people will choose to adopt these new technologies.

This is a prediction that is easy to make. As science-fiction author William Gibson keenly observed: "The future is already here, it's just not evenly distributed yet." Science fiction is a literature of ideas, and the best of the genre considers the consequences of the possible long before it becomes the plausible. Although early in development, and unevenly distributed, rudimentary neural prostheses work today. No fundamental barrier to this technology is set by physics, chemistry, or biology – the pace of progress will depend only on the rate at which we reduce our ignorance about how to build devices. Future demand is certain to be enormous, and deployment will soon be determined more by product-development timelines than by basic science. Yet every technology arrives with surprises. Just as today the "digital divide" of differential internet access is identified as a form of educational discrimination, will brain implants demolish or extend these barriers? What sorts of new social or economic classes will we incidentally construct as human augmentation becomes a matter of whether we can afford the latest upgrade? We are taking very real steps into the opening chapters of every Gibson novel yet written, and we must be ready to embrace those still brewing in his head.

When software meets wetware

Eventually, genetic engineering, regenerative medicine and information technology will converge. Bridging the gap between computation and living matter will bring both benefits and challenges, just as networking has complicated the security of banking, utilities and manufacturing. Which brings us again to Gibson and the opening pages of his debut novel, *Neuromancer*, wherein the protagonist discovers that his engineered, transplanted organs have been hacked to release a toxin unless he performs a task to receive an antidote. There are already many engineered organs in clinical trials, and when this future becomes evenly distributed we must be concerned about malware installed at the factory in our wetware. What if that wetware needs some sort of upgrade after it has been installed? Who is in charge of managing these updates, and do they have the authority to push out new code over a biological network, rather like smartphone software updates are delivered today? In other words, will we have to visit a doctor for tune-ups, or will biological code be spread in other ways, perhaps refreshing the original sense of "going viral"? Will we be able to refuse these updates? Ultimately, who will hold the "password" for our transplant, whatever that will mean in this context?

Again, as strange as it may seem, nothing in this scenario is prohibited by what we know of physics, biology or chemistry. As with the cortical modem, there is enormous demand for technologies that will reduce the burden of disease, improve the quality of life as we age and, ultimately, convert ageing from something with an inevitable endpoint into a managed, continuing process. Still, it will be a long road.

Both the cortical modem and regenerative medicine exemplify the human capacity to engage in engineering amid a sea of unknowns. We cannot yet design a brain, or anything that works even vaguely like a brain, because we have an incomplete understanding of how the constituent cells work, either individually or collectively. Nevertheless, we are vaulting over our ignorance to construct new capabilities for human physiology. We can now read and write just enough of the language of neurons to connect them directly to inorganic computers. We have just enough control over the behaviour of human cells to paste them together in useful shapes that, by mechanisms we do not fully understand, become functioning organs. This is an indication that the scope of biotechnology in coming decades will be limited not by what we know today of biological bits and pieces, but rather by how well we kluge together contraptions that just work. This is a modus operandi that has, historically, paid enormous dividends.

The basis of biotech's boom

Although we are still learning how to efficiently and safely repair and modify human bodies, including our genomes, we have been reading and writing genetic code of other organisms in the laboratory for decades. The demand for such technology is substantial. Appreciating the economic pull on biotechnology, despite its early stage, is crucial to understanding where we are headed. Commercial activity based on genetic modification has gradually, and quietly, become a major contributor to the American economy.

By 2012, US revenues from biotechnology had surpassed 2% of GDP (see Figure 3.1). Three major subsectors make up these revenues: biologics (ie, biopharmaceuticals), genetically modified crops and industrial biotechnology (such as fuels, enzymes and materials). If considered as an industry in itself, biotechnology contributed more to the US economy in 2012 than did mining (0.9%), utilities (1.5%) or the manufacturing of computer and electronic products (1.6%). Should the relative size of biotechnology come as a surprise, this is because it has been poorly measured. The contribution of semiconductors to the US economy was tracked by the Department of Commerce at least as early

FIG 3.1 **Lift-off** US biotech revenues, $ billion

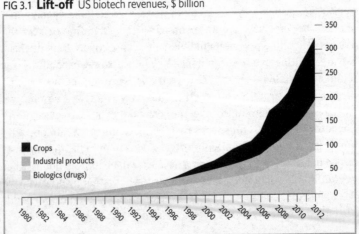

Sources: *Bioeconomy Capital; Nature Biotechnology*

FIG 3.2 **Carlson's curve v Moore's law**

DNA synthesis and sequencing productivity*, compared with microchip progress

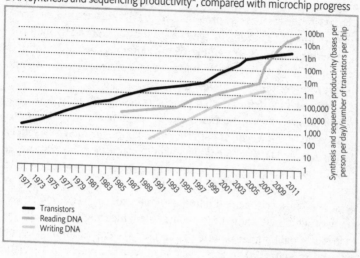

Synthesis and sequences productivity (bases per person per day)/number of transistors per chip

100bn
10bn
1bn
100m
10m
1m
100,000
10,000
1,000
100
10
1

1971 1973 1975 1977 1979 1981 1983 1985 1987 1989 1991 1993 1995 1997 1999 2001 2003 2005 2007 2009 2011

— Transistors
Reading DNA
Writing DNA

*Using commercially available instruments.
Source: Robert Carlson

as 1958, when it was less than 0.1% of GDP; as of 2016 there was still no official tracking of biotechnology. As a result, its economic impact has snuck up on us.

Revenue from biotechnology increasingly rests on the ability to read, alter and write genetic code one DNA base pair at a time. Automated instruments now directly interconvert electronic and biological instructions, a capability developed over the past 30 years. This timescale is important, because it is shorter than the time remaining to 2050, the end of the period considered by this book. In another 30 years, this technology will be inexpensive, pervasive and vastly more capable. Since 1985 costs for reading and writing DNA have plummeted, while the hourly throughput of instrumentation has improved at least exponentially, doubling every 18 months or so. In recent years, sequencing productivity has improved hyper-exponentially, driven by soaring demand to read the genetic instructions encoding people, pathogens, tumours, crops, pets and every naturally occurring organism that scientists can lay their hands on.

Once digitised, these gene sequences constitute a collection of instructions useful for building new genetic capabilities into production organisms, to date generally microbes and plants. This DNA may code for just one protein used as a pharmaceutical, or it may code for an entire enzymatic pathway that can produce any molecule now derived from a barrel of petroleum. After decades of learning to program biology, we are not limited in our designs to the genes we find in nature, or by their arrangement.

It is now possible to design genetic code according to particular functional specifications and then to integrate these instructions into a genome. However, as with the cortical modem and replacement tissues, we are starting to engineer genetic circuits without knowing how all the parts work. The most complex synthetic genetic circuits presently in commercial use consist of only 12 or so genes pasted into yeast genomes, which are themselves composed of over 5,000 genes, many of which we know little about. Bioengineering is presently an exercise in hacking an inherited, complex system without a decent instruction manual. Massive investment will be applied to alleviate that ignorance over the next 30 years.

We will spend those decades learning about how all the parts and systems underlying life fit together. These components are finite in number, function, and interaction, and there is little doubt that over time we will fully describe them. As our understanding improves, growing market demand will inevitably drive the refinement of engineering capabilities. What is the world going to look like when we actually know what we are doing?

The future, bioengineered

To develop some intuition about the future of biological engineering, consider the following exercise in reverse engineering. How would a Boeing 777 in operation have been viewed in 1892, a century before it took flight? In an era when automobiles were the big news, with horses and their manure still the norm in daily life, every aspect of a modern flying machine would have been a mystery. The materials and methods used in its construction, the engines and systems that keep it airborne, and the computation and complexity-management enabling the autopilot to handle 90% of take-offs and landings in all weather: all would have been considered completely implausible. Though clearly allowed by laws of physics (which have not changed) in 1892, the 777 was simply beyond contemporaneous imagination or engineering capabilities.

Then, over the course of 100 years, the bits and pieces that go into a 777 were improved, refined and combined into a functioning whole that today seems mundane. This infrastructure is now so mature, and so well integrated, that designers can sit at desktops and drive automated manufacturing lines half a world away.

Yet we still do not fully understand how airfoils sustain lift in turbulent airflow. Rather than basing our designs on a detailed physical description of flight, we make do with computational models of fluid dynamics informed by vast numbers of simulations. Ultimately, we trust the airworthiness of the 777 to these simulations. Nonetheless, the products that emerge from the aviation "design for manufacturing" system are so safe and so highly reproducible that they regularly lull us to sleep just after take-off. That we find modern aviation less than thrilling is actually an exciting clue to the future of biotechnology. While it may sound trite, the future lies in making biological manufacturing as boring as building and flying aircraft.

The transformation is already under way. A biodesign automation industry, analogous to the one enabling modern aviation, is now emerging from ambitious start-ups on multiple continents. Large pharmaceutical and industrial biotechnology companies, unable nimbly to redirect their own research-and-development efforts down this path, are the primary customers for this new approach. When design for manufacturing eventually becomes a mundane aspect of bioengineering, we will have access to a platform technology that can be used to build just about anything we see in nature. In that future, the boundaries of biotechnology will be expanded far beyond the limited descriptions of biological parts and processes that we work from today.

As we extend our ability to manipulate biology, our creativity, over-constrained by present experience, will gradually be set free. What will we build using biological components when we can transcend the imaginary limits imposed by current understanding? We can see hints of the future through another DARPA project, one that aims to use biology to change the way we manipulate inanimate matter.

Standard synthetic chemistry has provided a zoo of molecules that are the building blocks of the modern economy. Many products today are possible only through properties of molecules that are entirely of human manufacture. Whether through plastics, coatings or catalysts, synthetic chemistry literally transforms our world. But of all the materials we can theoretically imagine, synthetic chemistry can be used to manufacture only a fraction. Enzymes, however, can manage feats of chemistry that provide access to a much larger number of potential materials. DARPA wants to extend this capability and employ novel combinations of enzymes to build a thousand materials that have never existed before. Moreover, after a century of effort, we have learned enough biochemistry to start designing new enzymes, with new capabilities, that expand even further the accessible gamut of the materials spectrum.

Thus, even while we struggle accurately and comprehensively to describe biology as we find it, we are already using biology to escape the constraints of technologies developed before the 21st century. Beyond manufacturing novel materials, biological technologies are being eyed as important functional components of systems now produced from silicon and metal. In particular, biotechnology is likely to transform the way we store digital information.

From disks to DNA

The internet is expanding so rapidly that our need to archive data will soon outstrip existing technologies. If we continue down our current path, in coming decades we will need not only exponentially more magnetic tape, disk drives or flash memory, but exponentially more factories to produce these storage media, and exponentially more warehouses to store them. Even if this is technically feasible, it is economically implausible. Biology can provide a solution. DNA is by far the most sophisticated and densest information-storage medium we have ever encountered, exceeding by many times even the theoretical capacity of magnetic tape or solid-state storage.

A massive warehouse full of magnetic tapes might be replaced by an amount of DNA the size of a sugar cube. Moreover, whereas tape might last decades and paper might last millennia, we have found intact DNA in animal carcasses that have spent 750,000 years frozen in the

Canadian tundra. Consequently, there is a push to combine our ability to read and write DNA with our accelerating need for more long-term information storage. Encoding and retrieval of text, photos and video in DNA has already been demonstrated.

Governments and corporations alike have recognised the opportunity. Both are funding research to support scaling up the infrastructure to synthesise and sequence DNA at sufficient rates. In order to compete with a typical tape drive, a single "DNA drive" must be able to write and read the equivalent of approximately ten human genomes a minute, which is more than ten times the current global annual demand for synthetic DNA. The scale of the demand for DNA storage, and the price at which it must operate, will completely alter the economics of reading and writing genetic information, marginalising the use by existing multibillion-dollar biotech markets while at the same time massively expanding capabilities to reprogram life. This sort of pull on biotechnology from non-traditional applications will only increase with time.

Land of milk and biomoney

Consider the manufacturing potential of fermentation, a fusion of biology and process engineering. Beer brewing works technically and economically at all scales, from the millions of litres per year sluiced out by multinational giants down to the thousands finely crafted by the corner microbrewery. This industrial structure demonstrates that distributed biological manufacturing can compete against centralised manufacturing, rebutting the notion that economies of scale always favour the large. Moreover, integrated petroleum companies are viable only in capital amounts of tens of billions of dollars, whereas fermentation-based businesses are possible with merely thousands of dollars of infrastructure.

By reprogramming the biological portion of this manufacturing platform, we can aim the flexible whole at quite lucrative markets. Whereas beer is mostly water, worth at most a few dollars per litre, microbes can produce molecules worth tens of thousands of dollars per litre. Of the more than $105 billion contributed by industrial biotech to the US economy in 2012, at least $66 billion came from fermented biochemicals that are already displacing petrochemicals from global markets (this excludes bioethanol, which added only $10 billion to US GDP in 2012). As another indicator of demand, the pharmaceutical industry is switching from chemical synthesis of even small-molecule drugs such as antibiotics to biological synthesis, thereby saving money and reducing waste streams and carbon emissions. The rising demand for these renewable chemicals will be met increasingly by production systems comprising both biological and non-biological components.

Farmers have already discovered the advantages of hybrid approaches in the form of robotic milking machines for dairy operations. The cows and robots together comprise an integrated system with superior productivity and profitability. More than 25,000 such systems are in operation globally. Cows rapidly learn to visit the milking barn at times of their choosing, where their health and productivity are tracked by electronic tags embedded in their collars, a sort of "internet of cows". Importantly, the cows appear to benefit from this system, as they require fewer visits by veterinarians and produce more milk. The networked cows forage for food, process it on the hoof into a valuable substance and then autonomously deliver it to a centralised facility for collection.

The critical insight here is that, like beer brewing, dairy farms augmented with automation comprise an extremely productive, flexible and distributed manufacturing system. This integration is extending a multi-decadal trend in which overall milk production has nearly doubled while the size of the "dairy fleet", so to speak, has been reduced by half.

Now, imagine if those "surplus" cows could make fuels or chemicals instead of milk – the volume of this production would be equivalent to the US's renewable-fuels mandate in 2017, or approximately 17% of total US gasoline demand.

According to current plans, meeting that mandate will require an estimated $170 billion to build hundreds of presently hypothetical "integrated biorefineries". In contrast, the value today of the US dairy fleet is approximately $20 billion. If to this figure we added as much as $10 billion spent on sorting out how to engineer cows to make fuels and chemicals, we would still be ahead by comfortably more than $100 billion.

Yet development costs are likely to be even lower, because we do not have to go so far as bioengineering actual cows to deliver products to robotic milking barns. We already know that we can build fermentation systems containing engineered microbes that consume complex organic feedstocks and produce valuable chemicals, all the while outcompeting petrochemicals. And we can see an impending future of autonomous robots that employ wheels or legs for locomotion. The combination of these technologies will completely transform the way we manage resources and organise manufacturing.

Howdy, cowborg

Imagine robots equipped with bioprocessing modules slowly wandering around pastures, or even restored high-plains grasslands, consuming a variety of feedstocks, processing that material into products ranging from fuels and chemicals to pharmaceuticals, and then delivering it to collection facilities. The robots might look like cows, or simply like today's automated, satellite-guided harvesting equipment augmented with fermentation tanks. These hybrid "cowborgs" – in essence, mobile microbreweries – would be autonomous, distributed biomanufacturing platforms.

Whatever the final form of these hybrids, we will employ biological components, robots, or digital computers where each is best suited. The overarching message is not that the limits of biology will be expanded by computation, but that both kinds of technology will head off in new directions due to the impact of the other.

If this sounds like a fantastical vision, recall that this book encompasses developments spanning more than three decades into the future. Because the benefits of biotechnology are already creating massive demand, and because barriers to entry are falling steeply, in 30 years we should expect that our economy will rely heavily on hybrid devices that combine engineered biotic and abiotic parts.

It is difficult to predict specifically what artefacts constructed in that future will do or look like. But, looking ahead, the key theme is that the constraints of the past, let alone the present, will not apply. The future will be defined not by biology as we find it today, but rather as we will build it tomorrow.

Note

I would like to thank Rik Wehbring, Sarah Keller, Eric Carlson, Spencer Adler and Stephen Aldrich for fruitful conversations and hard questions.

Beyond Moore's law

Tim Cross

**Extraordinary advances in microprocessing power
have enabled a revolution in computing. In future, the
revolution will have to continue by other means**

IN 1971 INTEL, then an obscure firm in what would only later
come to be known as Silicon Valley, released a chip called the 4004.
It was the world's first commercially available microprocessor, which
meant it sported all the electronic circuits necessary for advanced
number-crunching in a single, tiny package. It was a marvel of its time,
built from 2,300 tiny transistors, each around 10,000 nanometres
(or billionths of a metre) across – about the size of a red blood cell. A
transistor is an electronic switch that, by flipping between "on" and
"off", provides a physical representation of the 1s and 0s that are the
fundamental particles of information.

In 2015 Intel, by then the world's leading chipmaker, with revenues
of more than $55 billion that year, released its "Skylake" chips. The firm
no longer publishes exact numbers, but the best guess is that they have
about 1.5 billion–2 billion transistors apiece. Spaced 14 nanometres
apart, each is so tiny as to be literally invisible, for they are more than
an order of magnitude smaller than the wavelengths of light that
humans use to see.

Everyone knows that modern computers are better than old ones.
But it is hard to convey just how much better, for no other consumer
technology has improved at anything approaching a similar pace. The
standard analogy is with cars: if the car from 1971 had improved at the
same rate as computer chips, then by 2015 new models would have top
speeds of about 420 million miles per hour. That is roughly two-thirds

the speed of light, or fast enough to drive round the world in less than a fifth of a second. If that is still too slow, then before the end of 2017 faster-than-light models that can go twice as fast again will begin arriving in the showrooms.

This blistering progress is a consequence of an observation first made in 1965 by one of Intel's founders, Gordon Moore. Moore noted that the number of components that could be crammed onto an integrated circuit was doubling every year. Later amended to every two years, "Moore's law" has become a self-fulfilling prophecy that sets the pace for the entire computing industry. Each year firms such as Intel and the Taiwan Semiconductor Manufacturing Company spend billions of dollars figuring out how to keep shrinking the components that go into computer chips. Along the way, Moore's law has helped to build a world in which chips are built in to everything from kettles to cars (which can, increasingly, drive themselves), where millions of people relax in virtual worlds, financial markets are played by algorithms and pundits worry that artificial intelligence will soon take all the jobs.

No more room at the bottom

But it is also a force that is nearly spent. Shrinking a chip's components gets harder each time you do it, and with modern transistors having features measured in mere dozens of atoms, engineers are simply running out of room. There have been roughly 22 ticks of Moore's law since the launch of the 4004 in 1971 through to mid-2016. For the law to hold until 2050 means there will have to be 17 more, in which case those engineers would have to figure out how to build computers from components smaller than an atom of hydrogen, the smallest element there is. That, as far as anyone knows, is impossible.

Yet business will kill Moore's law before physics does, for the benefits of shrinking transistors are not what they used to be. Moore's law was given teeth by a related phenomenon called "Dennard scaling" (named for Robert Dennard, an IBM engineer who first formalised the idea in 1974), which states that shrinking a chip's components makes that chip faster, less power-hungry and cheaper to produce. Chips with smaller components, in other words, are better chips, which is why the computing industry has been able to persuade consumers to shell

FIG 4.1 **Moore or less** Stuttering progress

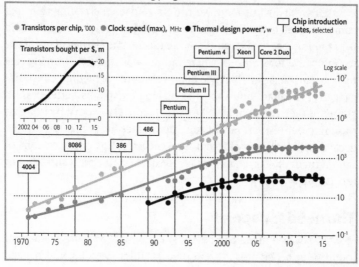

*Maximum safe power consumption.
Source: *The Economist*

out for the latest models every few years. But the old magic is fading. Shrinking chips no longer makes them faster or more efficient in the way that it used to (see Figure 4.1). At the same time, the rising cost of the ultra-sophisticated equipment needed to make the chips is eroding the financial gains. Moore's second law, more light-hearted than his first, states that the cost of a "foundry", as such factories are called, doubles every four years. A modern one leaves little change from $10 billion. Even for Intel, that is a lot of money.

The result is a consensus among Silicon Valley's experts that Moore's law is near its end. "From an economic standpoint, Moore's law is dead," says Linley Gwennap, who runs a Silicon Valley analysis firm. Dario Gil, IBM's head of research and development, is similarly frank: "I would say categorically that the future of computing cannot just be Moore's law any more." Bob Colwell, a former chip designer at Intel, thinks the industry may be able to get down to chips whose components are just 5 nanometres apart by the early 2020s – "but you'll struggle to persuade me that they'll get much further than that".

One of the most powerful technological forces of the past 50 years, in other words, will soon have run its course. The assumption that computers will carry on getting better and cheaper at breakneck speed is baked into people's ideas about the future. It underlies many of the technological forecasts elsewhere in this book, from self-driving cars to better artificial intelligence and ever more compelling consumer gadgetry. There are other ways of making computers better besides shrinking their components. The end of Moore's law does not mean that the computer revolution will stall. But it does mean that the coming decades will look very different from the preceding ones, for none of the alternatives is as reliable, or as repeatable, as the great shrinkage of the past half-century.

The need for speed

Moore's law has made computers smaller, transforming them from room-filling behemoths to svelte, pocket-filling slabs. It has also made them more frugal: a smartphone that packs more computing power than was available to entire nations in 1971 can last a day or more on a single battery charge. But its most famous effect has been to make computers faster. By 2050, when Moore's law will be ancient history, engineers will have to make use of a string of other tricks if they are to keep computers getting faster.

There are some easy wins. One is better programming. The breakneck pace of Moore's law has in the past left software firms with little time to streamline their products. The fact that their customers would be buying faster machines every few years weakened the incentive even further: the easiest way to speed up sluggish code might simply be to wait a year or two for hardware to catch up. As Moore's law winds down, the famously short product cycles of the computing industry may start to lengthen, giving programmers more time to polish their work.

Another is to design chips that trade general mathematical prowess for more specialised hardware. Modern chips are starting to feature specialised circuits designed to speed up common tasks, such as decompressing a film, performing the complex calculations required for encryption or drawing the complicated 3D graphics used in video games. As computers spread into all sorts of other products, such specialised silicon will be very useful. Self-driving cars, for instance, will increasingly make use of machine vision, in which computers learn to interpret images from the real world, classifying objects and extracting information, which is a computationally demanding task. Specialised circuitry will provide a significant boost.

However, for computing to continue to improve at the rate to which everyone has become accustomed, something more radical will be needed. One idea is to try to keep Moore's law going by moving it into the third dimension. Modern chips are essentially flat, but researchers are toying with chips that stack their components on top of each other. Even if the footprint of such chips stops shrinking, building up would allow their designers to keep cramming in more components, just as tower blocks can house more people in a given area than low-rise houses.

The first such devices are already coming to market: Samsung, a big South Korean microelectronics firm, sells hard drives whose memory chips are stacked in several layers. The technology holds huge promise. Modern computers mount their memory several centimetres from their processors. At silicon speeds a centimetre is a long way, meaning significant delays whenever new data need to be fetched. A 3D chip could eliminate that bottleneck by sandwiching layers of processing logic between layers of memory. IBM reckons that 3D chips could allow designers to shrink a supercomputer that currently fills a building to something the size of a shoebox.

But making it work will require some fundamental design changes. Modern chips already run hot, requiring beefy heatsinks and fans to keep them cool. A 3D chip would be even worse, for the surface area available to remove heat would grow much more slowly than the volume that generates it. For the same reason, there are problems with getting enough electricity and data into such a chip to keep it powered and fed with numbers to crunch. IBM's shoebox supercomputer would therefore require liquid cooling. Microscopic channels would be drilled into each chip, allowing cooling liquid to flow through. At the same time, the firm believes that the coolant can double as a power source. The idea is to use it as the electrolyte in a flow battery, in which electrolyte flows past fixed electrodes.

There are more exotic ideas, too. Quantum computing proposes to use the counter-intuitive rules of quantum mechanics to build machines that can solve certain types of mathematical problem far more quickly than any conventional computer, no matter how fast or high-tech (for many other problems, though, a quantum machine would offer no advantage). Their most famous application is cracking some cryptographic codes, but their most important use may be accurately simulating the quantum subtleties of chemistry, a problem that has thousands of uses in manufacturing and industry but which conventional machines find almost completely intractable.

A decade ago quantum computing was confined to speculative research within universities. These days several big firms are pouring money into the technology, including Microsoft, IBM and Google, all of which forecast that quantum chips should be available within the next decade or two (indeed, anyone who is interested can already play with one of IBM's quantum chips remotely, programming it via the internet). A Canadian firm called D-Wave already sells a limited quantum computer, which can perform just one mathematical function, though it is not yet clear whether that specific machine is really faster than a non-quantum model.

Like 3D chips, quantum computers need specialised care and feeding. For a quantum computer to work its internals must be sealed off from the outside world. Quantum computers must be chilled with liquid helium to within a hair's breadth of absolute zero, and protected by sophisticated shielding, for even the smallest pulse of heat or stray electromagnetic wave could ruin the delicate quantum states that such machines rely on.

Fading from sight

Each of these prospective improvements, though, is limited: either the gains are a one-off, or they apply only to certain sorts of calculations. The great strength of Moore's law was that it improved everything, every couple of years, with metronomic regularity. Progress in the future will be bittier, more unpredictable and more erratic. And, unlike the glory days, it is not clear how well any of this translates to consumer products. Few people would want a cryogenically cooled quantum PC or smartphone, after all. Ditto liquid cooling, which is heavy, messy and complicated. Even building specialised logic for a given task is worthwhile only if it will be regularly used.

But all three technologies will work well in data centres, where they will help to power another big trend of the next few decades. Traditionally, a computer has been a box on your desk or in your pocket. In the future the increasingly ubiquitous connectivity provided by the internet and the mobile-phone network will allow a great deal of computing power to be hidden away in data centres, with customers making use of it as and when they need it. In other words, computing will become a utility that is tapped on demand, like electricity or water today.

The ability to remove the hardware that does the computational heavy lifting from the hunk of plastic with which users actually interact – known as "cloud computing" – will be one of the most important ways for the industry to blunt the impact of the demise of Moore's law. Unlike a smartphone or a PC, which can only grow so large, data centres can be made more powerful simply by building them bigger. As the world's demand for computing continues to expand, an increasing proportion of it will take place in shadowy warehouses hundreds of miles from the users who are being served.

This is already beginning to happen. Take an app like Siri, Apple's voice-powered personal assistant. Decoding human speech and working out the intent behind an instruction such as "Siri, find me some Indian restaurants nearby" requires more computing power than an iPhone has available. Instead, the phone simply records its user's voice and forwards the information to a beefier computer in one of Apple's data centres. Once that remote computer has figured out an appropriate response, it sends the information back to the iPhone.

The same model can be applied to much more than just smartphones. Chips have already made their way into things not normally thought of as computers, from cars to medical implants to televisions and kettles, and the process is accelerating. Dubbed the "internet of things" (IoT), the idea is to embed computing into almost every conceivable object. Smart clothes will use a home network to tell a washing machine what settings to use; smart paving slabs will monitor pedestrian traffic in cities and give governments forensically detailed maps of air pollution. Once again, a glimpse of that future is visible already: engineers at firms such as Rolls-Royce can even now monitor dozens of performance indicators for individual jet engines in flight, for instance. Smart home hubs, which allow their owners to control everything from lighting to their kitchen appliances with a smartphone, have been popular among tech-savvy early adopters.

But for the IoT to reach its full potential will require some way to make sense of the torrents of data that billions of embedded chips will throw off. The IoT chips themselves will not be up to the task: the chip embedded in a smart paving slab, for instance, will have to be as cheap as possible, and very frugal with its power: since connecting individual paving stones to the electricity network is impractical, such chips will have to scavenge energy from heat, footfalls or even ambient electromagnetic radiation.

Mooreover

As Moore's law runs into the sand, then, the definition of "better" will change. Besides the avenues outlined above, many other possible paths look promising. Much effort is going into improving the energy efficiency of computers, for instance. This matters for several reasons: consumers want their smartphones to have longer battery life; the IoT will require computers to be deployed in places where mains power is not available; and the sheer amount of computing going on is already consuming something like 2% of the world's electricity generation.

User interfaces are another area ripe for improvement, for today's technology is ancient. Keyboards are a direct descendant of mechanical typewriters. The mouse was first demonstrated in 1968, as were the "graphical user interfaces", such as Windows or iOS, which have replaced the arcane text symbols of early computers with friendly icons and windows. CERN, Europe's particle-physics laboratory, pioneered touchscreens in the 1970s.

Siri may leave your phone and become omnipresent: artificial intelligence will (and cloud computing could) allow virtually any machine, no matter how individually feeble, to be controlled simply by talking to it. Samsung already makes a voice-controlled television. Technologies like gesture tracking and gaze tracking, currently being pioneered for virtual-reality video games, may also prove useful. Augmented reality (AR), a close cousin of virtual reality that involves laying computer-generated information over the top of the real world, will begin to blend the virtual and the real. Google may have sent its Glass AR headset back to the drawing board, but something very like it will probably find a use one day. And the firm is working on electronic contact lenses that could perform similar functions while being much less intrusive.

Moore's law cannot go on for ever. But as it fades, it will fade in importance. It mattered a lot when your computer was confined to a box on your desk, and when computers were too slow to perform many desirable tasks. It gave a gigantic global industry a master metronome, and a future without it will see computing progress become harder, more fitful and more irregular. But progress will still happen. The computer of 2050 will be a system of tiny chips embedded in everything from your kitchen counter to your car. Most of them will have access to vast amounts of computing power delivered wirelessly, through the internet, and you will interact with them by speaking to the room. Trillions of tiny chips will be scattered through every corner of the physical environment, making a world more comprehensible and more monitored than ever before. Moore's law may soon be over. The computing revolution is not.

5

Tech generations: the past as prologue

Ann Winblad

To get a sense of how things will change over the next three decades, consider how successive waves of technology have formed over the past three

IN THE AUTUMN OF 1985, about as many years back from today as this book looks ahead, Bill Gates and I took a long walk on an empty North Carolina beach. It was a decade after Bill had founded his software company as a partnership, and just a few months before (on March 13th 1986) Microsoft was to sell its shares publicly for the first time.

Microsoft was one of the early entrants in the "second wave" of computing technology: personal computers. But in 1985 few were aware of any new breaking waves in computing technology, let alone that we were already on the second one. The first, from the late 1950s through the 1970s, ushered in the mainframe and minicomputers. Bill and I, while both in our teens, had been swept into the software and technology industry during this first wave. Bill in high school and I in my freshman year in college had access to minicomputers: DEC (Digital Equipment Corporation) VAXes. I learned to program in Fortran on a DEC PDP-11. Bill was far ahead of me, programming on several minicomputers including the DEC PDP-10.

At that time computing power, on mainframes and minicomputers, either in large cooled rooms or via time-sharing, was invisible to most. Although it was just the first wave, it foreshadowed how the subsequent ones would break. Each brought stunning newcomers. What is initially visible above the surface is only part of it. The young wave extends down through the water column to the ocean floor, where the many new companies initially swim. The wave tilts and at some point curls over and forms a breaker as it approaches the shore. This is when the winners appear and the losers crash – only the leading innovators and companies ultimately come ashore.

The first wave

In the first wave Burroughs, UNIVAC, Control Data, Honeywell, RCA and General Electric would swim strongly. This initial group of mainframe manufacturers was referred to as "IBM and the Seven Dwarfs" and later as "IBM and the BUNCH" when General Electric and RCA left the business. The BUNCH would essentially disappear or become irrelevant – from a historical perspective, they were to be only momentarily visible at the surface as the wave crashed. IBM would come ashore as the big winner from this first wave, although today just a small share of IBM's revenue lingers from mainframes. On the minicomputer front, names like Apollo Computer, Data General, Wang Laboratories, Prime Computer and even the better-known DEC all remain mere historical footnotes.

In each wave a few core technology components or new business strategies emerge that leave technological footprints in the sands of time, frequently taking decades to show their full impact. Out of the first wave would come Moore's law (see Chapter 4). Gordon Moore's 1965 paper positing the annual doubling of processing power is still a driver of new waves. The waves of computing technology have pretty much kept pace with Moore's law, leading to astonishing advances in technology capabilities. Moore's law, whose rate was halved from annually to every two years in 1975, would set the cadence for all future waves. It would also determine much more than the number of transistors on a chip: it would define the pace of innovation in manufacturing, design and software.

On our beach walk Bill was pensive. He was not concerned that, far off the shore, the third wave of computing, which came to be called Web 1.0, was already forming. He had other things on his mind. Microsoft had ended its fiscal year in June 1985 with just over $140 million in revenue. This was a stunning amount for a young company in the new PC business. The nascent software industry was bubbling with competition. Many were listed in Microsoft's public-offering prospectus, including Lotus Software, Ashton-Tate, Software Publishing, Borland International and Digital Research. All were growing companies swimming under the surface of a wave that was at the time still far from closing in on the shore. As we walked Bill explained how he could do the maths to get Microsoft to $500 million in revenues, but struggled to see getting beyond that. No software company had even approached such a heady number.

Winners in the second wave

In this second wave battles raged among operating systems and in desktop software, among both hardware manufacturers and the emerging pure software companies. Apple would reach the shore as a new type of information-technology company. Microsoft's software competitors were among the minnows. Many of the other visible companies, including Lotus Software, would soon disappear entirely or be acquired as small divisions of larger companies. By 1990 this wave of computing technology had crashed to the shore with Microsoft and its proprietary software emerging as the big winner. Bill's concerns about the size and scale of his young company were quickly behind him. Microsoft reached $590 million in revenues in just two years from its public offering, and $1.1 billion in 1990.

Along with Microsoft, Oracle and its distributed database also emerged strongly. This second wave, compared with the first, was clearly more visible. The PC would reach hundreds of millions of desktops as shipments grew from 50,000 in 1975 to over 134 million in 2000. Software became an industry.

In every wave "prototypes" of the future would surface. Early on enthusiasts cheer and embrace these early entrants. In the era of the second wave Bill and I were carrying Motorola DynaTAC mobile phones – phones the size and nearly the weight of a brick. Motorola's popular StarTAC flip phone would not appear for another decade. We both lugged the 16-pound Macintosh Portable in its big square backpack as we travelled.

Bill shared one bigger, longer-term concern with me on that walk in 1985: would we humans become carbon-based pets by 2050? Later, in 1993, this idea was described as the "Singularity" by Vernor Vinge, a computer scientist and science-fiction writer. Vinge suggested that between 2005 and 2030 the acceleration of technology innovation would lead to machine intelligence that could match and surpass human intelligence. In fact, it would take almost 25 years from my walk with Bill for the seventh wave, artificial intelligence (AI), even to begin forming far offshore.

Faster still and faster: the third and fourth waves

Two waves of computing technology have reached the shore since the one that gave us Microsoft. Wave three, Web 1.0, washed in the internet as well as Amazon and Google. Wave four, Web 2.0, brought "the cloud" and mobile computing; Apple with the iPhone, Google with Android, Amazon with Amazon Web Services (AWS) were winners, and a new company, Facebook, came ashore strongly. The first cloud-based software companies arrived in this wave, most notably Salesforce.

FIG 5.1 **Wave power** The swelling of new technologies

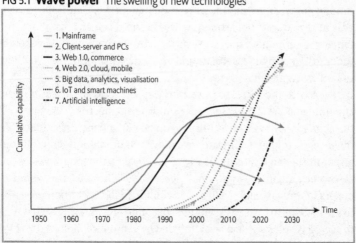

Sources: Accenture; World Economic Forum

Each wave of computing technology has been stronger, almost exponentially so, experiencing a multiplier effect from the previous ones. The pace of growth of new companies coming ashore has also accelerated dramatically. Whereas Microsoft took 15 years to reach $1 billion in revenue, Google, founded in 1998, reached $1 billion in five years and by year 15 surpassed $50 billion in revenue. Facebook crossed $1 billion in just over four years. Amazon, an online retailer which started in 1994 in wave three (Internet 1.0), reached its first $10 billion in revenue in 13 years; and AWS, the cloud segment of Amazon, a wave-four (Web 2.0) entrant, reached $10 billion in just ten years from its release in 2006.

Wave power and venture capitalists

Money also increasingly powered the waves, in the form of venture capital. Venture capital began in 1959 with the backing of Fairchild Semiconductor by what would later become the Rockefeller family-supported Venrock Associates. Independent venture-capital firms did not appear until the early 1970s, beginning with Kleiner, Perkins, Caufield & Byers and Sequoia Capital. Even then the amount invested was small.

It was not until 1978 that venture capital experienced its first major fundraising year, as the industry raised about $750 million. In that year the US Labour Department relaxed certain restrictions under the Employee Retirement Income Security Act, allowing corporate pension funds to invest in the asset class and providing a major source of money to venture capitalists.

Even with the increase in venture capital, software investing was still limited. Fear of the assets – the software engineers – walking out of the door at night, as well as the fledgling nature of business models in this new sector, kept software investing to $400 million–600 million a year in the late 1980s and early 1990s. In 1995 the total invested in software companies would finally exceed $1 billion. By 2015 venture dollars in software had swelled to $23 billion of the $58 billion invested in the US. This increased the number of companies entering each wave. In 1995, 435 software deals were funded by venture capitalists. By 2015 that number had increased to over 1,800. The winners in software also grew fast, both organically and by acquiring many other new companies. Microsoft's revenue reached $93 billion by 2015. Salesforce, a fourth-wave company, became the sixth-largest software company with $6 billion in revenue. Amazon, with $107 billion in revenue, and Google, with almost $75 billion, came to top the internet-company list.

How software was built and what drove innovation changed in the 1990s. In the third wave free or open-source software became a popular choice for website developers. Open-source software such as Linux decreased the cost of starting a new company. Such software was not only free but also supported and rapidly enhanced by a global community of developers. In 2000 Salesforce publicly released APIs (application program interfaces), sets of programming instructions and standards for accessing a web-based software application. No longer was proprietary software a winning strategy; software and the internet became open and "programmable" by all. Today over 15,000 APIs exist for every category of software.

Waves five and six: big data and the internet of things

The internet brought an explosion of data. So-called big data became so large and complex that traditional data-processing applications or platforms were inadequate for capturing all the information, sharing it, storing it and searching it, let alone for any predictive analytics. In 2006 a seminal piece of open-source code called Hadoop was released. Apache Hadoop, a free open-source piece of software, coupled with cloud-computing platforms, started the big-data wave with a system that enabled distributed parallel processing of huge amounts of data across inexpensive industry-standard servers that could scale almost without limit.

Not only will billions of people on thousands of applications be generating these data, but in the next ten years it is estimated that over 100 billion devices – each with dozens of sensors – will be connected to the internet. Upwards of a trillion sensors will shortly be gathering data everywhere from the "internet of things" (IoT), whether in wearable devices, autonomous cars, drones, satellites or cameras. As the big-data wave builds, so too does the IoT wave that follows it, as we move from the plain internet to the "internet of everything", with network connections between many billions of devices as well as billions of people. These two last waves, big data and IoT, have not yet reached shore as we see the seventh wave coming, the one Bill was most concerned about: AI.

Come in number seven

AI companies first showed up in the 1980s in the second wave. In the 1980s expert systems began appearing on PCs. Universities offered expert-system courses and many large companies applied the technology in their businesses. Venture capitalists also funded a few new entrants: Aion Corporation, Neuron Data, Intellicorp and Inference were the leading ones at the time. But by the early 1990s the term "expert system" and the AI companies in this wave had all but disappeared. None reached the shore.

As with other waves, enthusiasts are already confidently proclaiming that AI is now here, based on many prototypes of the future. Key building blocks for modern AI have arrived with each subsequent wave. AI technology itself has also been rethought since the 1990s. However, we are very early in the wave and not likely to reach Singularity within Vinge's time frame.

The AI era of expert systems morphed into a new field called "machine learning". This explores algorithms that can learn from data and make predictions. "Deep learning" is the newest branch: its algorithms are based on data generated by interactions of multiple layers of machine learning. The exponential increase in digitised data to feed learning systems, improvements in tools for the data, key open-source software and inexpensive cloud infrastructure have led to an explosion of innovation in modern AI.

Venture capitalists have again shown investing interest in AI. Early-stage venture capitalists place many of their investments – quietly naming some companies as merely "stealth" – as the waves are forming offshore, so many of the entrants are unknown, even as the investing dollars are known. Seventh-wave investing in modern AI probably began around 2010 (see Figure 5.2). AI start-ups received investments of $2.6 billion in 2015. Applications leveraging the AI-infrastructure

FIG 5.2 **The next big thing** Start-up investment in artificial intelligence

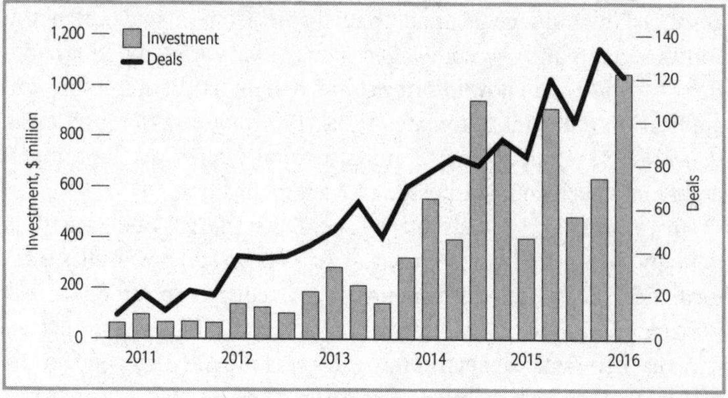

Source: CBInsights

tools received investments of $3.6 billion in 2015. IDC, a research firm, estimates that as of the end of 2015 only about 1% of all software applications had AI features, many of these the newly funded entrants. IDC also predicts that by 2020 the market for machine learning will reach $40 billion and that 60% of these applications will run on the platforms of Amazon, Google, IBM and Microsoft.

The IoT and its more tangible embodiments make an explosion of digital technologies and intelligence more obvious and interesting to the broader public. The IoT adds contextual awareness to everything. My Nest Learning Thermostat, now a Google product, is programmable and self-learning. It optimises the heating and cooling of my home and sends me satisfying messages of how I have conserved energy. It is based on a machine-learning algorithm. For the first few weeks I had to regulate the thermostat in order to provide the reference data set. Now it "thinks" it has learned my schedule and which temperature to use when. Using built-in sensors and my phone's location, it shifts into energy-saving mode when it realises I am not at home. My Rachio sprinkler controller intelligently analyses recent and upcoming weather and humidity levels, no longer requiring me to read weather forecasts and turn my sprinkler controller off. From security cameras to door locks to refrigerators, home devices all act smarter as they incorporate data and machine learning.

Arthur C. Clarke, a science-fiction writer, said: "Any sufficiently advanced technology is indistinguishable from magic." Both the software giants and new companies in this AI wave are competing in the rise of "bots" and in how to surprise and delight us. A bot is a software application that runs automated tasks. Typically, bots perform tasks that are both straightforward and structurally repetitive, at a much higher rate than would be possible for a human alone.

In a world of thousands of apps and now hundreds of devices, it is not hard to delight. I delight in my Amazon Echo and how well "Alexa" recognises my voice and responds to my requests to rapidly access massive amounts of data, as well as in the fact that it is now connected to many of my smarter devices through Amazon's APIs. I depend on Siri to send text messages while I drive. I marvel when Google anticipates my search requests with Google Now. I live in a world of too many apps and allow the bots to help me. I will eventually look back to today and will probably say that these bots were "prototypes for the future": in essence rapid search engines with great voice recognition. My expectations are high that sooner rather than later I will have a real conversation with Apple's Siri, Amazon's Alexa, Microsoft's Cortana or some break-out company's product. (Alexa already has a sense of humour: when I ask her to "open the pod bay doors", she answers, "I'm sorry Dave, I can't do that. I'm not Hal and we're not in space.")

While we await the shift to computers run by conversation and the ultimate Turing test – the test of AI, developed by Alan Turing in 1950, of a machine's ability to exhibit intelligent behaviour equivalent to, or indistinguishable from, that of a human – there is no question that we are making software smarter. A rapid, virtuous and competitive cycle of innovation has picked up invisible momentum as the seventh wave builds. This embedded intelligence has appeared broadly as software gets smarter: improved recommendation engines on your commerce apps, programmatic ad placement, medical diagnostics, outbound call-centre software, investment-portfolio selection, analysing and optimising risk, pricing products, routing transport.

Tasks, and indeed professions, that it was once thought could be done only by humans are increasingly done better in partnership with machines. Fed the right data, algorithms can schedule, analyse, decide, predict, diagnose and even write news stories, rapidly penetrating the realm of easily repeatable tasks, complementing us with intelligence applied to enormous datasets. The appetite for more and more data to feed the machine has for the first time led to the breakdown of isolated silos of data created by competition across companies in various industries. This collaboration could, in the not too distant future, lead to massive impacts in the sciences that affect our lives and health.

When the wave crashes

In the decades that it will take for the new winning companies in the AI wave to reach the shore, new battles will ensue. The competitive dynamics and areas of opportunity will change. The open and collaborative software supply chains may seem to be a distinct advantage for future winners, but in many ways this openness makes the competitive race faster and harder. Unlike Microsoft entering the second wave in a nascent software sector, companies entering the seventh wave will compete with many proven winners, including IBM, Microsoft, Google, Amazon, Facebook and others. And although venture capitalists will probably continue to place the majority of their funds in software companies, their sights will need to be reset to outpace these previous winners.

Already there are not-so-subtle changes in the target opportunities for both entrepreneurs and investors. Software investment for the most part has historically been focused on developing tools, platforms and applications to allow enterprises to achieve greater operating efficiency or scale. In many ways, software entrepreneurs and their venture investors have been the arms merchants to the enterprise. Yet Amazon, Netflix and more recently Uber and Airbnb stand out as companies attacking the enterprises' core businesses: retail, entertainment, transport and hospitality respectively. Software now defines much of the customer experience and has given the customer more power and transparency.

Uber, founded in 2009, is a fifth-wave, big-data company. For Uber, a world where everyone has in their pocket an always-connected, multi-core computer with geolocation allows the customer to order a car at any time. The same device enables drivers to sign on and start and stop work as they desire. Algorithms applied to many big-data sources

FIG 5.3 **Softsurge: the software supply chain**
Download requests for open-source components, billion

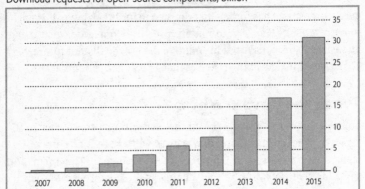

(ranging from weather, news and cultural events to other transport strikes) determine pricing and the best route. It does not require much imagination to see how AI will benefit Uber with greater insights for its business and engagement by its customers. Machine learning will allow Uber to better anticipate supply and demand to set its prices. Autonomous vehicles will complete its fleet. Natural language processing will bring more engagement and data with its customers. The story is similar for Airbnb in the hospitality industry.

Entrepreneurs and venture capitalists together have begun their march to unbundle all that can be digital in industry after industry, throwing down the gauntlet to global businesses in their quest to attack bigger opportunities. Venture-capital investment for new software companies unbundling just one industry, financial services, reached $13.8 billion in 2015, more than double the total invested in such "fintech" in 2014 and six times more than the funding deployed in 2011. Other industries follow the same pattern. Armies of software developers are working furiously to compete. The software supply chain, that is the open-source download requests from the industry's major source-code repository of open-source components to build code, Maven Central, illustrates the accelerating pace of development (see Figure 5.3).

Bill Gates was correct in his belief that software could build enormous value. But even he now says that the risk of AI software becoming "super-smart" is far into the future. The AI wave is very young. It is unlikely in this wave that a machine will ever match the creativity of a Gates, Steve Jobs or Mark Zuckerberg, or of the new entrepreneurs in the coming waves. However, there is now no doubt that the ability to harness the "internet of everything" and intelligently leverage massive data to serve and empower customers will be a critical battleground.

6

The great innovation debate

Ryan Avent

An argument rages over whether technology will in future bring anything like the surge in growth and productivity that it did in the past

IN THE SPACE OF JUST A FEW WEEKS in early 2016, one could just about see the dawn of a new and bright technological age creep over the horizon. In Europe, convoys of trucks crossed the continent while their drivers relaxed and handed control to computers for hours at a time. In the middle of the Atlantic Ocean, SpaceX, a private spaceflight company, landed a reusable rocket capable of delivering satellites into space on a computer-controlled barge. And in Seoul, AlphaGo, a powerful artificial-intelligence system built by Google, defeated the world's best player of Go – a game with far too many potential moves to be "solved" through brute-force calculation, as chess was.

Less than two decades into the new millennium, humanity is building new technologies with seemingly limitless practical applications. Yet even as the world watched these marvels in wonder, a deep pessimism permeated discussions about the future of growth. Around the same time Robert Gordon, an economist at Northwestern University, published an impressive book on the past and future of US productivity growth. In *The Rise and Fall of American Growth: The U.S. Standard of Living Since the Civil War* he argued that a great wave of innovation in the latter half of the 19th century transformed rich economies and provided the fuel for a century of rapid productivity growth. Electricity and automobiles, indoor plumbing and modern medicine laid the groundwork for the decades of profound change that created the modern world.

FIG 6.1 **A productivity puzzle** US labour productivity, annual % change

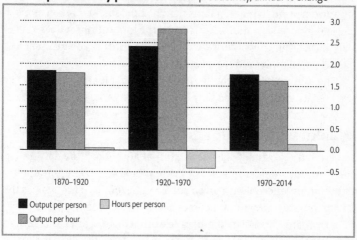

■ Output per person	▨ Hours per person	
▨ Output per hour		

Source: *The Rise and Fall of American Growth: The U.S. Standard of Living Since the Civil War,*
by Robert Gordon, 2016

Gordon does not see a repeat performance in the offing. The digital revolution, he argues, though important, is comparatively limited in its transformative potential. Decades of advance in information technology have not generated anything like the soaring growth in output per person, adjusted for inflation, that industrialised countries enjoyed in the mid 20th century (see Figure 6.1). Life expectancy is not growing as fast as it once did. The visions of a wildly prosperous future imagined in the 1950s and 1960s, based on robotics and rocketry and powerful computing, have failed to materialise. Gordon has only to point to the world around him to support his argument. The wonders of the digital revolution have grown in their power and capabilities, but the pay for many workers, adjusted for inflation, has grown at nothing like the rates observed a half-century ago. The societies of the rich world are angry and frustrated rather than satisfied and optimistic.

The difficult question to answer is whether technology will continue to disappoint. When machines converse with people in natural language and driverless cars deliver the precise goods people want to their home before they even realise they want them, will the world still feel as stuck in a rut as it does today? Gordon, and many others, say yes.

Digital technology, however nifty, cannot generate the same qualitative improvement in living standards that the fundamental innovations of the late 19th century did. What is more, the innovations it does deliver will be pushing growth forward in the face of stiff headwinds, such as ageing populations and soaring inequality.

The optimists, of whom there are also plenty, say no. They countenance patience. And they have the stronger case.

Weakness in numbers

Productivity pessimists have a big initial advantage: the data are on their side. Economists point to growth in productivity, or the amount of output produced with a given set of land, labour and capital, as the key to long-run growth in incomes and living standards. Productivity growth roared ahead in the decades immediately after the second world war across the rich world, but it decelerated sharply in the 1970s. In the late 1990s productivity growth rebounded, in the US especially, and many economists hailed the arrival, at long last, of the dividend from information technology. Yet by the mid-2000s the boomlet had petered out again; nary a recovery is in sight.

That, Gordon reckons, is all there was. The growth spurt of the tech boom represented the capture of gains from digitisation, personal computing and the internet, he says. As snazzy as more recent advances on those technologies have been, they are insufficient to raise productivity growth. Mobile technology and social networks do not much change humanity's ability to produce more with less; we were promised flying cars, to paraphrase Peter Thiel, a venture capitalist, but wound up instead with social networks. Driverless cars are not productivity boosters, since people can only be so productive when they are sitting in a car, whether or not they are driving.

Meanwhile, pessimists point out, the pace of innovation in computing, which helped sustain progress towards existing technologies, is slowing. For half a century engineers successfully kept up with the rule of thumb known as Moore's law (after Gordon Moore, a founder of Intel, a computing firm – see Chapter 4), which supposed that the number of transistors on a chip doubled once every two years or so. This impressive growth allowed computer-makers to shrink their products down from massively expensive, room-sized energy hogs to the supercomputers we now carry in our pockets. Unfortunately, Moore's law is now running out of steam.

All things considered, Gordon argues, the outlook for a technology-driven renaissance in the first half of the third millennium seems bleak. Are the prospects really so grim?

Some economists question whether the problem Gordon identifies is just a statistical illusion: measurements of the economy have not kept pace with technological change. The value produced in service and information industries, which account for ever more economic activity, is much harder to assess than the output of factories or farms. Many wondrous new digital goods are free, like Wikipedia or the suite of services provided by Google. What's more, increases in the value of consumption are more and more the result of rising quality or personalisation, like playlists from streaming services tailored to the listener's taste. Government statisticians struggle mightily to capture such nuances. Yet although mismeasurement is certainly responsible for some of the shortfall, number-crunchers reckon it is not the most important factor. Many of the same problems plagued government data collection in the late 1990s, at a time when measured productivity was soaring. And when researchers attempt to estimate the mismeasured value of new technologies, they come up with sums far lower than the estimated shortfall in overall productivity.

So Gordon's story is not a bad explanation for some of the economic disappointment of the past few decades. Yet it is almost certainly a poor guide to the next few. Indeed, the pessimists misread the nature of technological change in three ways.

Exponentially does it

First, they underestimate the cumulative effect of exponential improvement in computing power. Moore's law is indeed slowing down, yet its long lifespan has brought technology to the brink of extraordinary new developments. As Erik Brynjolfsson and Andrew McAfee, technology scholars at the Massachusetts Institute of Technology, have argued in several recent books, exponential growth processes are deceptive. They cite an old parable, in which a man invents the game of chess and brings it to the king for a reward. The man asks for rice, to be paid according to a particular pattern: one grain on the first square of the chessboard, two on the second, four on the third, and so on, doubling with each square. The king readily agrees, thinking the total must be a meagre one. Yet by the second half of the chessboard, the amounts are truly enormous – 4 billion grains on the first square of the second half – and each new square delivers a new payment of rice as big as all the ones that came before put together.

Similarly, the early doublings in computing power that occurred thanks to Moore's law delivered important but modest improvements in computing capability. But as time has passed, each generation provides a boost as big as everything that has gone before. Over the past decade, pessimists have repeatedly been surprised by the achievement of technological goals that only recently seemed to be years away. In the mid-2000s, self-driving cars looked well beyond the reach of available technology; just a few years later Google's self-driving cars were on city streets, and most manufacturers now sell vehicles with significant autonomous features. AlphaGo's victory likewise was achieved well ahead of schedule. Even if Moore's law slows further in future, each generation will provide a far more powerful contribution to computational progress than past doublings did.

A second reason for optimism is that Moore's law is anyway no longer a constraint on technological progress. Chipmakers are experimenting with new chip designs and materials in order to achieve performance improvements long after Moore's law ceases to apply. The massive amounts of cloud-computing power made available through companies like Amazon and Google mean that the speed of the chip on a user's desktop is increasingly irrelevant in determining the sorts of things a user can do. And improvements in computing power have also been augmented by advances in algorithms. AlphaGo's victory was made possible not simply through brute-force computation of possible game states, but through clever machine intelligence, which "thinks" through ways to beat opponents.

Taken together, these factors imply that there is much more room to increase the power and capabilities of thinking machines. They also suggest that advances in computing power do not simply nurture growth by helping people do things a bit faster with a somewhat smaller device from one year to the next. Rather, each generation of improvement pushes technology beyond new thresholds, opening up new possibilities to computer users.

Slowly, then all at once

If this is true, however, why hasn't technological advance led to more of a rise in growth, and how can we be sure such increases will be possible in future? The third, and strongest, reason for optimism is that it takes time to learn how to apply powerful new technologies.

Gordon is somewhat unfair to the digital revolution. He rightly credits big innovations such as electrification and automobiles with the long boom in output per person that the rich world enjoyed from the late 19th to the mid-20th century. He does not spend much time discussing a crucial point, however: that realising the potential of these innovations took a very long time indeed. The scientists experimenting with electricity had made many of the critical advances in the fundamentals by 1890. Yet productivity-boosting applications did not immediately arrive en masse. They appeared in dribs and drabs as firms found clever new ways to deploy electricity. Telegraphy appeared quite early on, for instance, but the broad electrification of homes and factories, and the productivity gains that followed, were not accomplished until much later.

Chad Syverson of the University of Chicago points out that productivity growth in the age of electricity was not uniformly rapid, but instead was disappointingly flat for long periods before accelerating upwards. He compares data on labour-productivity growth during the electrification age with the record so far in the IT era. The pattern is remarkably similar (see Figure 6.2).

The delay between the arrival of a technology and the full exploitation of its potential is mostly accounted for by the time needed to discover how best to use the new innovation and to rearrange the world accordingly. There were horseless carriages rattling around in the late 19th century, but it was a long time later that cars materially lifted growth. First, manufacturers had to figure out how to reduce their cost, governments had to amend regulations and invest in new forms of infrastructure, and firms had to experiment with new business

FIG 6.2 **Déjà vu** US labour-productivity patterns

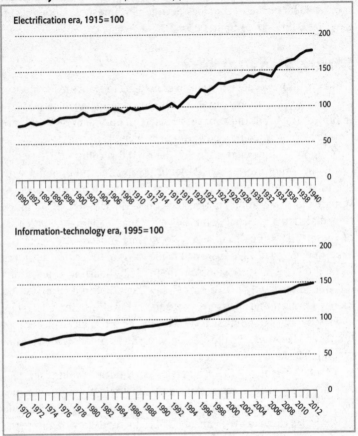

Source: Chad Syverson

models built around the automobile. People were still learning as of the latter decades of the 20th century, when the emergence of big-box stores raised productivity in the US retail sector.

These dynamics mean that productivity growth is always a reflection of the technological development that came before: between five and 15 years before on average, suggest economists Susanto Basu, of Boston College, and John Fernald, of the Federal Reserve Bank of San Francisco, but occasionally more. The productivity quickening of the late 1990s and early 2000s was mostly built on clever business-management software developed years before; web-based businesses contributed comparatively little. Similarly, it will be a while before things like driverless vehicles begin contributing to growth. The potential of machine learning cannot be written off by reference to today's weak growth.

Furthermore, when today's technologies begin affecting growth, it will be in ways that are difficult for us to imagine now. Karl Benz and Henry Ford could envisage the car as a better version of a horseless carriage, which could allow people to travel farther, faster, without relying on animals. But they could not foresee that cars would generate a dramatic change in the urban landscape, or that international trade would one day expand rapidly thanks to a system in which containers are lifted off ships and placed on trucks.

We are similarly unable to foresee the long-run applications and effects of driverless vehicles. They will almost certainly not be used in ways identical to cars driven by people. Instead, the fundamental nature of the car will change. Far fewer people might choose to own one, opting instead to hail a driverless car when they please. More efficient roads and reduced parking needs could change the structure of cities. But many more trips might well be taken by cars without any people at all, as households become increasingly accustomed to treating everything they need as available on demand: by asking aloud for a burger, for example, one might trigger a request from the household computer to a local restaurant, which would then place the meal in a tiny driverless vehicle to be ferried to the destination. Machines that are smart enough to navigate cars through traffic (and future machines will be capable of far more) can be deployed all across an economy: to conduct surgery, to tutor students in conversational speech, to run farms and manage energy systems, and more.

Computers are now small and cheap enough to go anywhere and be put in everything. Sophisticated machine intelligence will then allow them to manipulate the physical world in ways we cannot easily imagine. It is easy to imagine, however, that the transformation in society and economy resulting from this shift is as disruptive and powerful as the one that occurred alongside indoor plumbing, cars and electricity.

Social ties that bind

Some pessimism is justified, though not of the sort favoured by Gordon and his peers. If it is hard to imagine exactly how an AI-controlled house or car will change our lives in future, it is far easier to imagine the trouble society will have adjusting to the change. Promising technologies like driverless cars and drones are already running headlong into regulatory thickets. Governments are struggling to set rules for the collection and use of the massive amounts of personal data gathered by smartphones and other connected devices – even as the public grows nervous about governments' spying on those data. Before plumbing, electricity and cars could change the world, societies had to spend years investing in new infrastructure, experimenting with laws and regulations to work out who should own and operate such networks and under what terms, and evolving new cultural norms about what kind of behaviour was and was not appropriate. Humanity will go through that process again over the next few decades. This will slow the spread of new inventions and dampen their effect on the economy.

The trickiest adjustment of all, however, will be the management of the effect of these new technologies on labour markets and on the pay earned by workers. Indeed, it is possible that labour-market troubles are already having a serious detrimental effect on the use of new technologies and on productivity growth. Over the past few decades, wage growth for most workers in most rich countries has slowed by more than growth in the economy. At the same time, low levels of unemployment seem to be less effective at generating upward pressure on pay than was the case in the past (see Figure 6.3). Orthodox economists are used to thinking of productivity as a determinant of wages: as workers grow more productive, firms are able to pay them better. But some are also beginning to ask whether the link between low productivity and low wages may run in both directions.

Low pay allows firms to employ workers profitably in marginal jobs and to continue to use workers even though robots or software could replace them. Investments in automated checkout machines, for example, are less attractive when there are lots of cheap humans around. Some economists, such as João Paulo Pessoa and John Van Reenen of the London School of Economics, reckon that low UK wages,

FIG 6.3 **Wage gauge**
Change in annual real hourly wage growth between 2000–07 and 2007–14, %

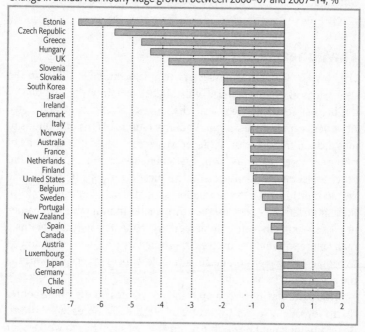

Source: OECD Employment Outlook 2015

which tumbled during the Great Recession, help to account for weak productivity growth during the subsequent recovery, since firms felt less pressure to economise. Similarly, abundant, cheap labour may help to explain how the US economy has managed to produce the unusual combination of soaring employment and weak wage growth in recent years.

As technology becomes more powerful, employers are able to find ways to replace workers and boost their bottom lines. Yet workers have to eat and usually seek other work when kicked out of their old jobs. As the glut of labour competing for jobs grows, wages stagnate or even fall. Low wages eventually make it attractive for firms to take on workers to perform low-productivity tasks. They also discourage new investments which might otherwise be made to save on expensive labour: in new automating systems or machine-learning programs to tackle tasks which can be done effectively enough by people provided they do not cost too much.

The upheaval ahead

The ways in which businesses and whole economies adjust to the technological changes ahead will almost certainly present one of the biggest social and political challenges between now and 2050. Driverless cars and trucks could quickly eliminate tens of millions of jobs around the rich world. Clever AI systems could displace tens of millions more, beginning with customer-service representatives and office assistants, then moving on to education and medicine, finance and accounting. Though some people should benefit fantastically from these innovations – because they own shares in profitable companies or possess skills that are complementary to the new machine brains – many more will find themselves threatened with displacement, forced to compete with many others for available work or to accept low pay as the price of retaining a job.

This pattern will leave people and economies on the whole poorer than they ought to be. Unfortunately, there are no easy solutions. Governments might begin to pay larger wage subsidies to workers, or even introduce unconditional basic-income payments to all citizens. But such payments will be expensive and will require heavy taxes on those getting rich from new technologies. Even if those picking up the bill consent, society as a whole might struggle to adapt to a world in which working is optional.

Governments might instead provide make-work jobs to the underemployed, but that would be expensive and wasteful. Or societies might simply become far more unequal than they are now, as technology creates a mass underclass of subservient service labour.

There is a precedent for this kind of difficult social adjustment. Early in the industrial era, the explosive growth of factory employment outstripped society's capacity to cope. Workers flooded into slums in cities without the infrastructure needed to provide clean water, or decent housing, or management of refuse and waste. The horrible living conditions that this brought about killed millions of workers. Those who survived earned meagre wages. Loss of work meant the risk of fatal poverty. Only after years of labour organising, social unrest, political reform and, in some cases, revolution did social institutions evolve in ways that facilitated the broad sharing of the gains from growth. These changes, which allowed workers to live longer, healthier lives, to get more education, and to save and invest more, also raised the capacity of the economy to grow by using new technologies.

Part of the reason growth in productivity and output has proved disappointing so far is the collision of new digital technologies with 19th- and 20th-century social institutions. In the absence of new reforms and investments, economies will continue to operate with vast reservoirs of underemployed less-skilled workers. These workers will hold down wages and discourage the use of clever new robots and thinking machines. If, in coming decades, society finds ways to allow workers to be more choosy in seeking where they work and how long to spend on the job, then firms might have an incentive to make better use of both technology and human labour. This could bring back the productivity growth of the good old decades of the 20th century, and make life far better for everyone.

PART 2

Megatech by sector

7

Farming tomorrow

Geoffrey Carr

How a planet of 10 billion people will feed itself

ON A BRIGHT SEPTEMBER DAY in 2050, Farmer Giles of Ham wakes up to a welcome sound: the shepherd's song from Beethoven's *Pastoral Symphony*. The harvesting app on his phone is telling him conditions are perfect to get in the crop from three of his ten barley fields. The others, it suggests, will be ready over the course of the next few days. The app will notify him when.

He and his wife are far from home on a weekend city break, but no matter. Last night they dined on *loup de mer*, fresh that afternoon, the menu claimed, from the pelagic shoaling tanks of the local Oceans Apart aquafab. The fish were served with a side of vegetables from Altitude, a chain of vertical farms whose slogan is "Food from the city, for the city".

He wipes the sleep from his eyes. It is just a question of reviewing the app's reasoning and pressing the "agree" button. The software behind it, mounted on a cloud computer who knows where, will then schedule the robot combine harvester he shares with four of his neighbours, all of whom also have several fields at peak readiness. Unfortunately, there is a clash. The app indicates that, collectively, one more field needs harvesting than the machine has time to manage. It could, of course, work through the night, since it has access via the cloud to precise topographical maps of all five neighbouring farms, so does not really need to see where it is going. But it would need to run along public roads to get between the farms, and the law will not permit that during the hours of darkness.

So someone will have to give way and, being in a good mood, he decides it will be him. One of his fields can wait until tomorrow. He has no doubt that, in the complex diplomacy of neighbourly farming, the favour will be repaid when he needs it.

The barley itself, a fertiliser-free strain that fixes its own nitrogen using live-in bacteria in its roots, has been genetically optimised both for his fields and for its ultimate destination, the local brewery. Since he is on excellent terms with the head brewer, he is looking forward to a barrel of the "Old and Nasty" his crop will end up in appearing mysteriously outside his kitchen door. The spent grain from the harvest will also return, to help give flavour to the drove of virus-proofed pigs that he keeps in semi-free range for optimal growth. He has shares in the Muscle Factory, of course, having cannily bought them in the initial public offering (IPO). But there will always be a premium for flesh straight off the bone.

The Muscle Factory and its competitors have, nevertheless, revolutionised things. Watching the urban trendies, divided between the no-Frankenfooders and the animal-welfarites, tearing each other apart over the matter of industrially grown meat had been amusing, but he had been sure the welfarites would win, and they had. A good thing, too. The factory farms of the 20th century were disgusting enterprises. Making animal-free meat for the mass market in a real factory was surely the way forward. He already had his eye on the IPO of Milkmade, which hoped to do the same thing to dairy farming.

*

This, or something like it, is one vision technologists have in mind for agriculture's future. Another, though, is of a rice-planter in rural Asia, or a rondavel owner in central Africa, who now farms mainly to subsist, joining the cash-crop economy with plants whose quality and yield have also been revolutionised by gene editing and genomic selection – and even by app-driven advice distributed over the mobile-phone network about when to plant and when to reap.

Both Farmer Giles and his Asian and African contemporaries will thus be the beneficiaries of technological advances whose outlines can already be dimly perceived, and whose antecedents go back two and a half centuries to the seed drill, the crop rotation and the scientific

FIG 7.1 **What's on the world's menu**
Daily calories per person by type of food

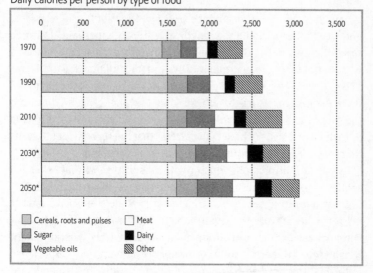

*Forecast.
Source: FAO

stock and crop breeding which kick-started the agricultural revolution in Britain in the mid-18th century. The results of what will, by 2050, have been three centuries of continuous technological improvement in the field of agriculture are that a world which could support fewer than a billion human beings when the revolution started will be able to feed 10 billion. The Food and Agriculture Organisation, the UN body that specialises in these things, envisages a further rise in calories per person in the more populous planet of 2050 (see Figure 7.1).

You say you want a revolution

One certainty about the farms of 2050 is that they will be even more mechanised, automated and factory-like than those of today. Jethro Tull's horse-drawn seed drill, which planted at regular intervals and buried the seed in the ground, is ancestor to all sorts of mechanical equipment. By 2050 this will include robot tractors and their associated paraphernalia, robot harvesters of the sort Farmer Giles shares with his neighbours, and drones and even satellites to monitor how a crop is doing.

Much of the irrigation (and the associated distribution of fertilisers and weed killers) will be the province of grids of pipes laid alongside rows of crops, rather than wasteful sprinkler systems. Soil monitors will control these pipes automatically. And for stuff that does still have to be sprayed – for example, pesticides and fungicides that need to fall upon the foliage, rather than the ground, to work – information from the drones and satellites will guide robots specially designed for the job, which will also be fitted with cameras that can recognise any weeds that have escaped attention and with lasers to zap them.

Moreover, farmers' relationship with the soil itself will have changed a lot by 2050. It has long been understood that soil is as much a product of what lives in it as of what minerals it is made from. This is what distinguishes it from the regolith that coats the moon and Mars. But better knowledge of the micro-organisms that are soils' most abundant inhabitants will permit soil improvements of equivalent importance to those introduced with crop rotation, by "Turnip" Townshend, and synthetic nitrogen-based fertiliser, by Fritz Haber.

Seeding soil with bacteria that fix nitrogen from the atmosphere and liberate chemically unavailable phosphorus will become routine. This will reduce the need for fertilisers of the sort Haber's artificial nitrogen-fixing process made possible. Most intriguing of all, though, will be better understanding of the relationship between crop plants and fungi. Many plants seem to engage in symbiotic relations with fungi, and these are only now starting to be unravelled. By 2050, such relations should be well enough understood for agronomists to exploit them.

What is less clear is how different the crops of 2050 will be from those of today. This is because the way new crops are created is about to change. Genetic engineering, which was supposed to help consumers as much as it did farmers, got off to a bad start in the 1990s. The companies engaged in it saw no fundamental difference between moving genes from, say, bacteria into maize or soya beans and tweaking those crops' genomes by irradiating them with radioactivity or applying DNA-altering chemicals. If anything, they assumed genetic engineering was better, as the process was less random. Aided by various hysterical lobby groups, however, much of the public came to a different conclusion. Though insect-proof and weed-killer-proof crops did well in many places, further genetic modification, for example to increase the nutritional value of crops, did not seem worth risking the investment.

That may change as a new generation of precise gene-editing tools are applied to crops. This time, seed companies' PR departments will get their retaliations in first, explaining to the public the value of what is being done, and that it is no more "Frankensteinish" than more established ways of breeding crops. Farmer Giles's nitrogen-fixing barley is an example of what might be done with such technology. But if public acceptance is there, things could go much further. The nutritional content of oilseed crops might be improved – for example, by adding much-prized omega-3 oils to species that currently major on the less-valued omega-6 versions of these molecules. Fruits might be altered to create new flavours or to enhance existing ones. The range of vegetables available to rich-world customers might be extended by tweaking currently non-commercial tropical crops so that they were amenable for sale to a mass market.

The real prize for 2050, though, would be to turbo-boost photosynthesis, causing crops to grow faster. Even now, researchers are working on an early form of this by trying to add what is called C4 photosynthesis to species which, at the moment, employ the more primitive and less efficient C3 version of the process. But this might be just the beginning. Many photosynthetic pathways unknown to plants exist in micro-organisms. Some may be ripe for transfer into crops. If enough consumers will accept that, the next two or three decades could see a Wild West of innovation in the improvement of yields. Fears that the human population would outstrip its ability to feed itself would evaporate.

Rus in urbe

All these thoughts, however, are fundamentally extensions of the way agriculture is done now. But as Farmer Giles's city sojourn suggests, some parts of agriculture in 2050 will be completely new.

The least novel of these new ways of doing business will be urban vegetable factories. Such factories will resemble in function, though not in form, the market gardens which supplied cities with fresh fruit and vegetables before the advent of mechanised transport and supermarkets. Produce will be sold, and usually consumed, on the day it is plucked. But urban vegetable factories will not be gardens open to the vagaries of the sun and the rain – and nor will they be the giant greenhouses of modern horticulture. Instead, they will be windowless buildings in which not only water and nutrients are controlled precisely, but also illumination. Their lighting will have its spectral composition adjusted to match exactly that used by chlorophyll, so that no photons are wasted.

Of intermediate novelty will be urban fish farms. Fish farming has been one of the greatest successes of the late 20th and early 21st centuries. The amount of fish protein produced in 2015 exceeded the amount of beef produced (see Figure 7.2). But these farms were mostly freshwater ponds or caged-off areas of arms of the sea, such as fjords. Urban fish farming will bring the oceans inland by providing closed-cycle systems in buildings to raise fertilised eggs to adult fish, and then use some of these fish to create the next generation. Moreover, once this process has been mastered for species that are already farmed, the road will be open to tame new ones, such as tuna, in a repeat, at sea, of the terrestrial farming revolution of the Neolithic, when most of what are now regarded as farmyard animals were domesticated for the benefit of mankind.

This process could change diets a lot. Fish are such efficient convertors of feed into flesh (far more so than mammals, because fish are cold-blooded and mammals warm-blooded) that it is possible to imagine that, by 2050, they will be the dominant form of animal protein. But it is also possible to imagine that they will not, and this might be because of the most novel of the new forms of agriculture that will be around then: true factory farming. If this technology comes to fruition, edible animal products will be grown from cell culture without the need for actual animals to be involved.

By 2050 it is likely that steak and milk, at least, will be manufactured in bulk. So will hens' eggs – though probably without shells, for industrial use rather than sale in the shops. And, for the more adventurous, organs such as liver and kidneys may be grown without

FIG 7.2 **Catching up** World farmed-fish and beef production, tonnes, million

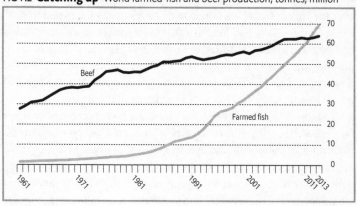

Source: FAO

the intervention of animals. As Farmer Giles observes, the evolution of this industry will probably cause a clash between two groups of idealists: those opposed to "synthetic" foods of any sort and those opposed to animal husbandry, particularly of the non-free-range variety. But the animal-welfare brigade is likely to win that one. Who could resist, for example, the idea of *foie gras* that could not possibly be accused of being the product of cruel practices?

The past, as a guide to the future

As always, though, the spread of technological change will both govern and be governed by social change. Farmers have always been conservative types and consumers, after a burst of willingness to accept the industrialisation and commodification of food after the second world war, also seem to have become conservative, at least in parts of the rich world. How consumer tastes will evolve is unpredictable. But, for some farmers at least, the social changes of the past may prove a guide to those of the future.

In particular, in poorer parts of the world, such as Africa, smallholdings whose primary purpose is to feed their owners will be consolidated into commercial farms whose primary purpose is to feed the market. This will involve decollectivising land ownership in places where collectivity is still the norm, and favouring the enterprising over the conventional – all things that have already happened in rich-world farming, often so long ago that their consequences, both good and bad, are the stuff of history books. But it is this change in land-holding patterns that will permit the deployment in such places of new technologies imagined in this chapter, and also of a good deal of existing technology that has yet to be rolled out there.

The consequence will be higher yields and higher incomes for those farmers who remain, while the surplus labour pushed off the land will find more valuable employment in cities. If that goes well, by 2050 African estate agents will perhaps have managed a trick pulled off by their European counterparts 100 years earlier. Those wily businessmen rebranded the thatch-roofed, mud-walled cottages once occupied by agricultural labourers as the most desirable of second homes, and city dwellers flocked to buy them. Why not a similar fate for rondavels on the shores of Lake Victoria or in the shadow of the Virunga mountains?

8

Health care meets patient power

Gianrico Farrugia

Rapid advances in medicine will both empower people and stretch society's ability to keep pace

MEDICINE HAS ALWAYS been a marriage of art and science, but technology, like never before, has been driving the advances in health care, as well as being its disruptive force. Historically, the life cycle of technology in medicine has been longer than that in other industries, but now it is rapidly accelerating. Amid all the upheaval ahead between now and 2050, one thing is clear. Patients, not technology, will be driving change in the future. This means that health care will more closely resemble other industries, where the patient is seen as a customer.

We need only go back 15 years to appreciate the unpredictability of technological advances on health care. Think about the technologies that were introduced then, and their impact on health care today.

The announcement in 2000 that the human genome was sequenced at first spawned a variety of technology companies vying to dominate the clinical-sequencing business. Most of them quickly went from boom to bust because the size of the dataset was too big and knowledge about the clinical use of the data was lacking. Yet what was abandoned by most life-science companies ten years ago is now one of the most thriving areas in medical research and health-care technology, one that has captured the attention of many governments. We may think that these sweeping changes came suddenly, but in reality the seeds were sown long before the value became apparent. Now, as commercialisation catches up with the potential of the science, the value becomes apparent. Similarly, we can use emerging knowledge of scientific advances both within and outside the field of medicine to predict the major technological developments in health care in the next decades.

Enter big data and artificial intelligence

Physicians have long been stymied by a lack of sufficient data when making a diagnosis. They now face the opposite issue: a risk of a cognitive overload. Until recently we used one X-ray to diagnose abdominal pain. Today, we use magnetic resonance imaging (MRI) technology that generates hundreds of images in the same amount of time. After initial attempts to use digital technology for diagnosis failed, doctors and software companies moved to computer-aided diagnostic tools with computer algorithms to guide interpretation of the data, where physicians would still have a final say. Even though natural-language processing has been around since the 1950s, its importance to health care is a recent phenomenon. We have now reached the point where our datasets are large enough, machine learning is sophisticated enough and the sums being invested are great enough (an estimated $8.5 billion globally in 2015) that it is increasingly evident health care will be disrupted by artificial intelligence (AI). We will see a new field emerging for specialised applications to summarise past history, determine the likelihood of positive testing outcomes and diagnoses, interpret tests, aggregate disparate data and personalise follow-up with patients. Tasks that are now assigned to humans, from the mundane monitoring of vital signs in intensive-care units to reading diagnostic images and performing the most complex surgery, will increasingly be carried out by learning machines.

This path is identical to that taken by the automobile industry, which went from cruise control to adaptive cruise control, and is now moving to driverless cars. The years ahead will bring stronger partnerships between life scientists and software engineers to create health-care technology that is not only useful but also aligned with the way physicians think. This will stimulate growth in the marketplace, which will produce ever more sophisticated AI tools to outperform physicians in certain cognitive tasks.

The rise of regenerative medicine and biotherapeutics

In the span of just a decade, regenerative medicine went from an emerging but promising field to becoming a reality. The body has incredible, built-in abilities to heal itself, and we are just beginning to harness this power. One type of therapy, called autologous therapy, involves regenerating a patient's cells in the lab and then returning them to the body. The regenerated cells then act as an engine and drive restorative function and healing. These advances will bring the initial promise of stem-cell therapy to life for a variety of diverse diseases.

There will be steady growth in natural products that are bioengineered, as subsets of diseases that respond to a particular biotherapeutic product are identified, and as the clinical utility of regenerative medicine increases. The commercial production of pharmaceutical-grade stem cells is an emerging need that will drive established and new companies to invest in the area. Today most biotherapeutics are non-cellular (monoclonal antibodies, for example, or growth factors); this will change if the initial promise of stem-cell therapy for disparate diseases such as joint pain, heart failure, stroke, ALS (amyotrophic lateral sclerosis, or motor neurone disease) and even diabetes and spinal-cord injury translates into successful large-scale clinical trials. The need to have on-demand personalised stem cells of known number, purity and stage of differentiation will give rise to a whole new industry.

As a resource derived from patients, a biotherapeutic product brings together the patient, the provider and the manufacturer. It significantly blurs the lines between production/manufacture (the traditional stronghold of companies) and delivery (the traditional stronghold of hospitals). This will require new types of business lines that do not exist today, with traditional manufacturing industries merging with biological providers. Using patient-derived resources on a large scale brings complexities that current suppliers are unable to meet. For clinical-grade regenerative-medicine products to achieve the scale required to change the health of populations, we need to create a whole new supply chain that integrates each part of the process, maintains the same quality from production to delivery, and can be individualised rapidly. Therefore, strong opportunities exist for all aspects of the supply chain, and these are unlikely to be filled by a single type of company. Niche areas will in all probability develop that specialise in "scaffolding" material, on-demand stem-cell production, instrumentation and much else.

Significant advances are also in prospect in immune-related therapies and vaccinology. We have seen the benefits of immune therapies for certain cancers and rheumatology, but the area will broaden to cardiovascular diseases, auto-immune diseases and many others. Vaccines remain a mainstay for the prevention of infectious diseases such as hepatitis B, polio and influenza, and they have eradicated smallpox. With some notable exceptions, less success has been achieved in the use of vaccines to treat active diseases, whether infective or oncological. This should change in the coming years with the merging of therapeutic vaccines and the current preventative ones. There are currently more than 1,000 active vaccine clinical trials with highly disparate targets.

One emerging domain is the use of vaccines to dampen selective parts of the immune system for diseases associated with auto-immunity, including type 1 diabetes. We already have vaccines designed to prevent cancer by targeting an infectious agent such as human papilloma virus. However, many cancer vaccines that initially showed promise have not passed final testing, decreasing enthusiasm and investment. A new crop of advanced cancer-vaccine trials now under way for a variety of cancers will be pivotal for the future of this field. There is also interest in a new breed of vaccines that are personalised based on genomic sequencing of tumours. Although advances in therapeutic vaccinology will tend to dominate media reports, a large market will remain for preventative vaccines, notably against infectious agents with increasing antibiotic resistance such as tuberculosis and respiratory infections.

High time for greater data integration

Because of legacy systems and heavy regulation, information gathered during encounters between medical providers and patients has been largely kept separate from data gathered by the patient through non-medically-approved devices (such as most wearable devices). For most consumers this is not acceptable, and very different from what we experience in the non-health-care part of our lives. The next generation of electronic medical records will be able to handle disparate data much better than current ones, but there still will be a big gap.

This creates a need. Data aggregators and technology companies built around taking the data and inputting relevant bits into the medical-health record will see a growing market for their services. We will see the emergence of technology companies that will serve the well person, the patient and the health-care provider by summarising large word- and number-based datasets, enabling and recommending decisions drawn from both data-delineated and non-data-delineated material. Information gathered by consumers through their own devices will increase in quality and quantity and approach clinical-grade, making it relevant to medical care and therefore creating a need for it to be incorporated in the electronic medical records. Wearable devices already provide 24-hour data points. This too will create a need: to use technology to separate consumer-obtained spurious data from meaningful information that informs rather than confuses the management of wellness and diseases.

The same issues that apply to the connected person also apply to the connected home. Technology will increasingly enable us to avoid trips to the medical-health provider, as our homes become an extension of ourselves, collecting information on our health and keeping us healthy. The medical provider will use our homes as the doctor's office and have access to data collected by the home, obviating the need for most visits. We will need the same technology that was developed for wearable devices to curate the interaction and ensure a focus on our needs rather than decisions made by technology alone.

An age of individualised medicine and "-omics"

The announcement in 2000 that the human genome was sequenced, followed by the completion of the sequence five years later, opened up the field of genomics (the aggregate of all our genes and DNA). It also gave rise to several new "-omic" fields, including pharmacogenomics (the interaction between drugs and our genes), epigenomics (the study of changes in gene expression that are not due to changes in the DNA) and proteomics (the large-scale study of proteins, their structure and function, and changes with time and illness). The commercial and health-care promise of genomics has been throttled by a lack of understanding about which of the tens of thousands of variants each of us carries predisposes us to disease, and about what combination of variants is additive or subtractive when it comes to health. In the early days the cost of sequencing was also a limiting factor. Crossing the $200 mark for a reliable sequenced genome (compared with $1,000 today), with some parts of the genome sequenced deeper than others, will remove that barrier. Our understanding of variants will continue to be a significant impediment to widespread use of new "-omic" testing and therapies, yet the robustness of rapidly increasing available databases suggests this will change over a five-year horizon.

Laboratory testing will markedly accelerate the use of next-generation sequencing to replace current methodologies such as neonatal screening, fluorescence in situ hybridisation (FISH) and a wide variety of genetic tests. In the diagnosis of rare diseases, the use of whole exome or genome sequencing will shift from being a last resort to first-line testing, broadening the market. Studies are emerging suggesting that for these rare diseases, whole exome or genome sequencing arrives at the answer faster and cheaper. With increased access to mass spectrometry, protein arrays and nanotechnology, as well as proteomic-based testing, we will expand disease insights downstream from genomic information, and in so doing may also increase our understanding of dynamic disease states. Although the application of the field of microfluidics to laboratory testing has paused, the technology will continue to make its way into labs, offering opportunities to decentralise them and creating point-of-care testing opportunities such as smart connected-home testing.

Pharmacogenomics will be an area of rapid growth, both for current laboratory-testing companies and for start-ups providing direct-to-consumer offerings. Of all the "-omic" areas this has the most utility data as well as existing delivery channels, and it is likely to expand rapidly. Over 150 medications (13% of the total) now have genomic information in their labelling. As price points settle, and as physicians and pharmacists get more comfortable with interpreting the data and applying them to prescription choices and dosage adjustments, both extraction of pharmacogenomics information from whole-genome data and targeted sequencing of specific genes will emerge as viable platforms. Still to be resolved is the full integration of pharmacogenomics gene sequencing with current software for electronic medical records and for prescribing drugs.

Targeted therapies (targeting the diseased molecule or cell rather than all cells) will dominate drug development. Although most people associate targeted therapy with cancer, the field is much broader than this: 43% of all drugs in the pipeline are targeted-therapy ones, according to the US Food and Drug Administration (FDA). In 2014, 20% of FDA approvals were for targeted therapies; in 2015 the figure rose to 28%. Growth in this area is due to the positive feedback loop of next-generation sequencing. The identification of targets leads to the development of targeted therapies, which then requires knowledge of affected targets, in turn driving further target discoveries.

Epigenomic changes are involved in a wide variety of chronic diseases – including metabolic diseases such as diabetes, obesity, heart disease and cancer – which are potentially reversible. Hence the strong interest in this field. Awareness is growing of the potential reversibility of epigenomic changes that drive disease causation. Epigenetic changes occur through a wide variety of mechanisms, most of which have only recently been discovered. We will see increasing attention to the development of DNA methylation-inhibiting drugs and drugs that target each mechanism, such as bromodomain inhibitors, histone acetyltransferase inhibitors, histone deacetylase inhibitors, histone methylation inhibitors and protein methyltransferase inhibitors. In addition, we will likely see the combination of epigenomic drugs with immunomodulators as standard chemotherapy regimens.

Molecular imaging in the mix

Advanced imaging modalities, such as computed tomography (CT) scans and MRI, are now 40 years old. Although massive improvements have been made and will continue to come in the core techniques, the addition of molecular imaging to these technologies and to newer technologies will dominate the next phase of imaging.

Molecular imaging is the combination of an imaging modality such as radiation, ultrasound, magnetism (MRI) or light with a mechanism to target a cell or components of cells such as specific molecules. In health care the term is typically applied to the use of imaging probes that target a specific molecule or pathway, which can be then visualised from outside the body. Positron emission tomography (PET) is the best-known molecular-imaging technique and relies on positron-emitting isotopes. Combinational imaging modalities will grow in the years ahead and dominate technology advances. It will become increasingly common to see a combination of the molecular targeting afforded by PET with the image quality of CT and MRI scans (PET-MRI and PET-CT). The use of newer contrast agents with higher specificity will also spread.

Hyper-ethical questions

This is an attempt to provide a glimpse into broad areas where technology will transform the future of medicine in the foreseeable future. We will still need new, disruptive medical technologies to complement these areas – especially with the rising tide of disability and chronic disease in our ageing population.

We have good reasons to be excited about the future, but it is not without risk. We are already living in a world where the velocity of innovation is accelerated by interconnectivity. Apps and mobile devices as delivery and consumption platforms have become mainstream in health care. Use of mobile devices as health-care products that provide medical-quality data will require further establishment of standards, but will increasingly enable testing to be decentralised and democratised.

The gap between the pace of innovation – occurring at an unprecedented rate – and both the regulatory environment and human adaptation is growing. How do we keep up, and make decisions on access and affordability? Who decides what is in the best interest of the patient? Thanks to telemedicine we are able to deliver better services more quickly and to faraway places, yet pandemics spread faster and farther as well.

With issues such as these, ethics needs to be the driver. We have a responsibility to ensure that patients' needs are not subordinated to technology, and that the quest for knowledge does not become Faustian – but benefits humanity instead.

9

Energy technology: the rise of the renewables

Anne Schukat

A great shift in energy consumption is on the horizon, thanks to advances in solar and wind power, as well as in the ways to store it

AFTER THE INDUSTRIAL REVOLUTION, the world came to depend on fossil fuels as its main sources of energy. This helped to create extraordinary economic growth, great advances in living standards and prosperity for many. Unfortunately, there is a downside. The burning of fossil fuels releases enormous amounts of pollutants and carbon dioxide into the atmosphere. For more than a century, humankind has been relying on an energy source that is dirty, finite and changing the climate irreversibly.

Over the next decades, however, a great shift will take place: away from fossil fuels. Technology improvements and cost reductions are advancing remarkably fast, in particular for solar and wind power. Their share in power generation could climb from 5% today to 30% by 2040, even if support from subsidies were phased out after 2020 (Figure 9.1 shows the similar picture for power capacity). In addition, batteries are getting better and cheaper, and may serve as an enabler for electric cars, as well as for absorbing more energy from renewables into the grid.

Unmistakable signs of this transformation are already here. According to the International Energy Agency (IEA), electricity generated by renewables accounted for around 90% of new power generation in 2015, with wind alone producing more than half. Similarly, Michael Liebreich, founder of Bloomberg New Energy

FIG 9.1 **Going greener**
Global installed power capacity, share of total

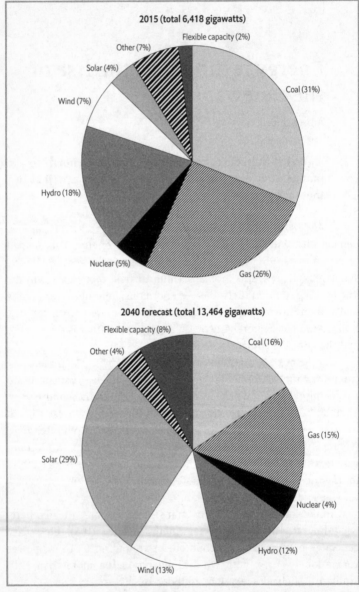

2015 (total 6,418 gigawatts)

Flexible capacity (2%)
Other (7%)
Solar (4%)
Wind (7%)
Hydro (18%)
Nuclear (5%)
Gas (26%)
Coal (31%)

2040 forecast (total 13,464 gigawatts)

Flexible capacity (8%)
Other (4%)
Solar (29%)
Wind (13%)
Hydro (12%)
Nuclear (4%)
Gas (15%)
Coal (16%)

Source: Bloomberg New Energy Finance, *New Energy Outlook 2016*

FIG 9.2 **Change of power** US energy consumption, % share

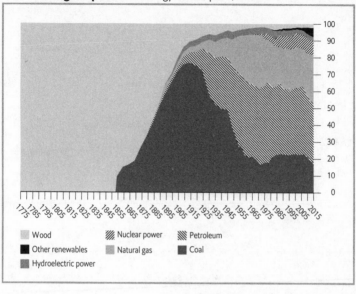

Source: US Energy Information Administration

Finance, a research firm, says the world is now adding more generating capacity to the electricity grid from renewables than from fossil fuels. Germany and California, two trailblazers, produce around 30% of their electricity from renewables.

Fossil fuels will not suddenly disappear; rather, their use will diminish over time. But the history of energy use is one of decades-long yet dramatic shifts in energy sources, from wood to coal to oil and gas (see Figure 9.2). And since energy production and use are responsible for two-thirds of greenhouse-gas emissions, the speed and scope of the next transition will determine whether global warming can be kept to a minimum.

Let the sun shine in

Solar power has come a long way since Bell Laboratories unveiled the first practical solar cell in 1954. Since then, the conversion efficiency of turning sunlight into electricity has nearly quadrupled, from 6% to 23%

FIG 9.3 **Sunny days** The price learning curve of solar power

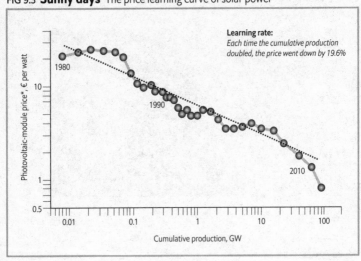

Learning rate:
Each time the cumulative production doubled, the price went down by 19.6%

*Inflation-adjusted.
Source: Fraunhofer Institute for Solar Energy Systems

for today's best silicon-based panels. Meanwhile, the cost for modules has dropped from almost $300 a watt in the 1950s to around 60 cents today (see Figure 9.3). As a result, solar power is already cost-competitive with fossil fuels in some places without government subsidies.

Solar power now generates about 1% of global electricity. While that may seem small, the photovoltaics (PV) industry has been expanding at a torrid pace. PV installations rose at a compound annual growth rate of 44% between 2000 and 2014. Since 2012, more solar PV capacity has been added worldwide than in all the previous years combined.

Solar cells are made from light-absorbing materials that convert sunlight into electricity. Today most often that is silicon, a brittle material that needs to be encapsulated and mounted in a rigid frame to make it durable. This limits the panels' deployment to rooftops or large installations in fields. Still, according to *The Future of Solar Energy*, a report from the Massachusetts Institute of Technology (MIT), today's silicon-based solar-cell technology is sufficiently good that it could be widely scaled up by 2050 to achieve large reductions in carbon emissions, even without major technological advances.

However, the report also concludes that technologies being developed now could potentially be easier and cheaper to manufacture, and could be deployed in a variety of forms, with conversion efficiencies similar to silicon. The new solar-cell materials could be deposited in thinner layers, on flexible substrates, and thus be more lightweight and easier to install. In addition, they could be made from transparent materials that absorb light not visible to the human eye and blend into any environment. Vladimir Bulović, co-chair of the report and professor of emerging technology at MIT, says the new technologies may have the ability to generate power on top of any surface.

If so, solar cells may start to find their way into a much wider array of applications. Over the next decades, as these technologies migrate from the lab to the marketplace, they may first appear in small electronic consumer items, on top of windows as transparent films, and later as part of fabrics, including curtains or clothing.

Both silicon and the emerging thin-film solar cells are made from abundant materials that can be produced at large volume, so wide-scale deployment would be no problem. According to MIT's calculations, the area required to supply 100% of projected US electricity demand in 2050 with currently available silicon-based technology would equate to around 0.4% of the country's land mass, or roughly half the size of West Virginia. However, deploying only the most efficient panels in particularly sunny parts of the US could help reduce that area by about two-thirds.

Blowin' in the wind

Similar to the sun, wind is a widely available, carbon-free and renewable source of energy. Wind turbines currently provide around 4% of global electricity. The cost of wind power has fallen from as much as 30 cents per kilowatt-hour in the 1980s to as low as 3 cents per kilowatt-hour now.

Whereas early wind turbines were short and small, with capacities of mere tens of kilowatts, today's machines are much bigger and reach higher, with typical capacities of around 2.5 megawatts and hub heights that range from 80 to 120 metres. The advantage of the taller towers is that they can access higher wind speeds at higher altitudes. Meanwhile, larger rotors with longer blades can sweep wider areas and collect more energy from the same location.

These continuing improvements are opening up vast new areas for wind development, says Edgar DeMeo, a managing consultant for the *Wind Vision* report from the US Department of Energy (DOE). According to the report, which was published in 2015 and looks at wind power's potential in the US to 2050, the next generation of turbines could add nearly 720,000 square miles (1.9 million square kilometres) to the country's developable land for wind and nearly triple the area that was accessible with older turbine technology from 2008.

Today's wind industry is dominated by traditional three-bladed turbines, but different designs are under development, including two-bladed machines as well as oscillating poles without any blades. The best-funded effort may be California-based Makani, which is developing airborne wind turbines that use a tether to connect and transmit energy to a ground station. The company's propeller-based "energy kites" operate on the same aerodynamic principles as conventional turbines but can potentially reach altitudes of up to 310 metres, or about twice their height, while using significantly less material.

In 2013 Makani was bought by Google. It is planning to test a 600-kilowatt prototype in Hawaii and is working with local pilots, as well as the US Federal Aviation Administration, to enhance the kite's visibility. But challenging the dominance of traditional turbines will not be easy. Today's machines have the advantage of a 30-year development history with clearly understood design criteria, says Paul Veers, chief engineer at the US National Wind Technology Centre. Current turbine technology is also still evolving. As rotors and towers get larger and land-based transport is becoming a challenge, companies are looking into assembling or manufacturing components on site. According to Jose Zayas, director of the DOE's Wind Energy Technologies Office, the 3D printing of blade moulds is now moving from the drawing board into prototype designs. The process should be faster than making blades from conventional moulds and may reduce their cost by 5%. Researchers are also investigating ways to optimise the layout and operation of wind plants, which should increase their productivity by around 5% at little additional cost.

For years mainstream energy forecasts have badly underestimated the rate of growth of both solar and wind power. For example, the IEA's *World Energy Outlook 2008* predicted that solar power would supply 1% of global electricity in 2030 – a point that was actually reached in 2015, 15 years early.

Solar power and wind power have one drawback, however: they are both intermittent. The sun doesn't shine at night and the wind doesn't always blow. It has been a learning process for utilities to integrate variable renewables into the grid. Today sophisticated forecasting methods anticipate when clouds cover the sun or the wind picks up, and grid operators balance supply and demand at much shorter intervals. A highly interconnected grid can be used to transfer excess energy over long distances where it is needed most. And if there is a shortage of sun or wind power, so-called natural-gas peaker plants can be fired up quickly to ramp up production.

But not all grids are equally robust or interconnected. And some places simply don't have the ability to send electricity over long distances or quickly fire up extra capacity when needed. Another solution to smooth out the spikes and troughs of intermittent renewables is to store surplus energy for later use – something that will become vital as the share of solar and wind in power production increases.

Power to the people

Although numerous storage technologies exist, they tend to be expensive. The commonest and lowest-tech option is pumped hydro, which involves pumping water uphill to a higher-level reservoir, and using it to spin a turbine when it is released and rushes back downhill. Another approach involves compressing air and confining it in large tanks or underground caverns. When needed, the air is released to spin a turbine and generate energy.

Different types of rechargeable batteries can also be used for storage. Batteries are made up of electrochemical cells, which use chemical reactions to generate electricity. The lithium-ion variety has become increasingly popular. Because it is relatively light and can pack a lot of energy into a small space, it can now be found in applications ranging from portable electronic devices to electric vehicles; scientists reckon they will be able to come up with further improvements to the battery's design and components, which could double its current energy density.

Another promising technology for grid applications is the flow battery. It consists of a set of tanks that hold two types of liquids and a separate electrochemical cell. When the two liquids are pumped through the cell, ions pass through a membrane from one fluid to another, while a proportional number of electrons make the journey through an outside circuit. Because flow batteries store their energy in the liquid electrolytes, their energy density is determined by the size of their storage tanks. At the moment, such batteries do not sell in large numbers. But if they could be made from cheaper and less toxic materials, they might become more widely available and could be used to provide many hours of storage.

California dreamin'

The world's installed storage capacity is tiny compared with its power-generation capacity. But that will change over the coming decades. California, which has a mandate to generate 50% of its electricity with renewables by 2030, requires its three largest investor-owned utilities to add 1.3 gigawatts of storage to the grid by 2020. Southern California Edison (SCE), a large utility that serves about 15 million people in Central and Southern California, has already procured several hundred megawatts of storage. AES Energy Storage, a subsidiary of AES, an energy giant, is installing a huge 100-megawatt lithium-ion battery system for the utility, which will be able to provide fast and flexible power for up to four hours. SCE is also working with Silicon Valley-based Stem, a company that combines small, modular lithium-ion battery storage with smart software to cut electricity costs for businesses and provide services to the grid. Its contract with SCE calls for installing 85 megawatts of distributed storage at around 1,000 of the utility's customers over a ten-year period.

By 2020, most new storage will be distributed and deployed "behind the meter" – not only at businesses but also in homes, says Ravi Manghani, director of energy storage at GTM Research. This is also what Christian Metzger, an expert on energy storage at RWE, a large German utility, is predicting. He expects that over the coming decades the combined distributed storage capacity of utility customers all over Germany will become large enough to supply all sorts of services to the grid, which will make building expensive large-scale storage projects unnecessary. Only after 2050, when renewables are expected to supply 80% or more of Germany's electricity, will additional long-term storage be needed, Metzger says.

For now, the technology of choice for new energy-storage systems worldwide is the lithium-ion battery, which accounted for more than 85% of deployed storage capacity in 2015, according to Navigant, a research firm. Tesla Motors, a California carmaker, is building a $5 billion "Gigafactory" in Nevada together with its battery supplier, Panasonic, to fulfil the rising demand for its cars, as well as its modular energy-storage system for homes and businesses, dubbed the "Powerwall" and "Powerpack", respectively. Other large manufacturers of lithium-ion batteries are also ramping up production.

All this should lead to much lower costs for battery packs, thanks to increased economies of scale, vertical integration and other efficiencies. According to a report from Bloomberg New Energy Finance, battery-pack costs for electric vehicles (EVs) may plummet from the current $350 per kilowatt-hour to below $120 by 2030. This should lead to a price point that would make EVs competitive with conventional cars without any subsidies. By 2040, the report says, 35% of all new global car sales could be EVs and plug-in hybrids.

As the share of renewables in the power sector increases, the energy to charge EVs will come from increasingly clean sources. Furthermore, EV owners will be able to offer up their car batteries for grid services to lower their electricity bills. The grid will become cleaner, more interconnected and distributed. Residential as well as commercial customers will be able not only to store energy in batteries themselves, but also to sell excess energy to others.

Long time coming

Although renewable and related technologies will have the most dramatic impact between now and 2050, other energy technologies will continue to evolve. Like renewables, nuclear power also generates electricity without releasing climate-changing gases. Nuclear fission involves splitting heavy atoms, such as uranium, into lighter ones and generating energy in the process. The first nuclear power plants started operating in the 1950s. Around 450 reactors are in use worldwide, providing about 11% of global electricity. Thus far nuclear power has avoided the release of about two years' worth of global CO_2 emissions at current rates, the IEA estimates.

In 2011 an earthquake and subsequent tsunami caused a series of meltdowns at the Fukushima power station in Japan. Although the released radiation did not kill anyone, it displaced more than 150,000 people. Public worries about further accidents, as well as concern over managing the radioactive waste, have made it harder to build more plants, especially in Western democracies.

Most of the 60 or so new reactors now under construction are in countries like China, India and Russia, where regulatory obstacles, and thus costs, are lower. An ageing cohort of around 200 nuclear reactors, located mostly in the US, Europe, Russia and Japan, may well be retired in the coming decades. As a result, the IEA predicts that the overall share of nuclear power in the electricity sector may grow little by 2040, if at all.

A different kind of nuclear power – the fusion of atoms – could potentially provide a much safer and nearly limitless power stream, without highly radioactive waste or the threat of a nuclear meltdown. During nuclear fusion, which is the process that powers the sun and other stars, lighter atoms, such as hydrogen, combine under intense heat and pressure to form heavier atoms, like helium, while releasing enormous amounts of energy. Since the 1950s, governments around the world have poured billions of dollars into developing the technology, and scientists back then predicted that working reactors would be built within a couple of decades. But replicating fusion on Earth has been more challenging than expected, and the original prediction has become a running joke, with the target appearing to be perpetually 20–30 years away.

Most recently, the internationally funded ITER (Latin for "The Way"), the biggest and most ambitious fusion project to date, has been riddled with delays. Originally slated to switch on in 2016, the massive reactor complex under construction in France is billions of dollars over budget and years behind schedule, with operation now estimated to start in about a decade. Yet many scientists in the field still see it as the best prospect to obtain fusion's holy grail: a reactor that produces much more energy than it consumes. At the moment, that goal is still far away. The 1997 world record for fusion power still stands at 16 megawatts – and required 24 megawatts of power input.

A handful of private companies have also taken on the problem. They reckon that they can make fusion work sooner and cheaper. Each one has a different solution to the fundamental problem of how to control and maintain the fuel particles (also known as plasma) at extremely high temperatures to facilitate fusion reactions. California-based Tri Alpha Energy has attracted hundreds of millions of dollars of investment, including capital from Microsoft's co-founder, Paul Allen. Its scheme involves high-energy particle beams to help heat and stabilise the plasma. Other firms, such as General Fusion in Canada and Helion Energy near Seattle, have also received money from high-profile investors, including Amazon's CEO, Jeff Bezos, and PayPal's co-founder, Peter Thiel, respectively. Whereas General Fusion uses a piston-based design to compress and heat the fuel particles, Helion prefers pulsed magnetic fields.

What these companies have in common is a belief that they can make fusion work in about 5–10 years. Some experts, however, are unimpressed. "Making such claims will undermine our credibility," says Steven Cowley, director of the UK's Culham Centre for Fusion Energy, who estimates that commercial reactors are still 30–40 years away.

Coal miner's blues

Meanwhile, there is no doubt that fossil fuels will stay with us for decades, even though their use will decline over time. And much could be done to make them cleaner and less detrimental to the planet while they are still in use.

At the moment, about one-third of new coal-fired plants being built and two-thirds of existing plants employ "subcritical" technology, with efficiencies of around 35%, whereas modern plants can reach 45% or higher. All coal plants can be equipped with filters, scrubbers or other controls that remove or reduce air pollutants – but, according to the IEA, this is often not the case.

Even less common is carbon capture and storage (CCS), the process of removing carbon dioxide from a plant's exhaust stream, and either storing it underground or recycling it for use in other industrial processes or products. At the moment only about 15 big CCS projects are in operation worldwide, whereas around 1,500 coal plants are planned or under construction. The first commercial-scale CCS system for a power plant, located in Canada at Boundary Dam, cost more than $1 billion to build. Since it was switched on in 2014, it has experienced technical problems and even shutdowns, and fallen far short of its target to capture 90% of its CO_2 emissions.

Proponents of CCS say new projects coming online will be less expensive and operate more smoothly based on lessons learned at Boundary Dam. But costs for such systems remain prohibitive. The pipeline of new CCS projects is shrinking: some 40 projects have been either put on hold or cancelled.

Compared with coal, natural gas releases less carbon dioxide and fewer pollutants while burning. In the US, the practice of hydraulic fracturing (known as fracking) of sedimentary rocks has opened up access to the country's vast shale-gas reserves and helped slash its coal use. But this type of fuel switching will remain mostly a US phenomenon, argues Seb Henbest, lead author of Bloomberg New Energy Finance's 2016 *New Energy Outlook*. In much of the rest of the world natural gas is shipped or piped in, which adds to its cost and will limit its growth over the next decades, he says. Developing countries may instead opt to add cheap coal and increasingly cheap renewables to their electricity portfolios. Public concerns about fracking-induced earthquakes, chemicals used in the process and the release of methane, a potent greenhouse gas, are also limiting shale exploration elsewhere.

The times they are a-changin'

By 2050, the Earth's population will be around 9.7 billion, compared with 7.4 billion today. Final energy demand will grow, especially in cities in emerging economies. According to the IEA's *Energy Technology Perspectives 2016*, new buildings equivalent to 40% of the world's existing structures will be added to urban areas in those countries by 2050, and passenger travel in cities will nearly double by mid-century.

But higher energy demand and better standards of living need not mean higher emissions. According to the IEA, new buildings could be equipped with highly efficient heating, cooling and lighting systems, as well as appliances. Reliance on public transport and electric cars could reduce CO_2 emissions as well as air pollution, particularly if they draw power from clean sources. Solar installations on cities' rooftops, for example, could generate a third of their electricity needs by 2050.

The exact scope and speed of the coming transformation is still uncertain. India, for example, has the ambitious plan to bring electricity to its 240 million citizens who are currently without power. Its goal is to do so in part by adding wind and solar installations, but it is also ramping up domestic coal production. China, meanwhile, is starting on a different path. In December 2015 it announced a three-year moratorium on new coalmines. It is also leading the world in zero-carbon technology investments, including wind, solar and nuclear power, with an average construction time for new nuclear plants of only 5.5 years. By 2040, Bloomberg New Energy Finance projects emissions from the Chinese power sector will fall by 5%, whereas India's could treble.

Globally, more than $430 trillion will be invested in the energy system in the coming decades, estimates the IEA. Based on the agency's calculations, investing an additional $12 trillion in low-carbon technologies by 2050 – roughly 3% more – could keep global warming to around 2 degrees Celsius, and improve air quality at the same time.

Technology is radically changing the energy outlook, forcing a rethink of what were until recently common assumptions about the resource constraints of the future. In place of energy scarcity, the prospect is for an age of energy efficiency and plenty. And plentiful energy need not mean more emissions and a more polluted planet. On the contrary, with sufficient investment in smart technologies, it could be a cleaner one.

Note

I would like to acknowledge the generous help I received from many people while preparing this chapter. Besides those named in the text, particular thanks go to the following: Menahem Anderman, Kamel

Ben Naceur, John Benner, Karen Butterfield, Sandy Butterfield, John Carrington, Katherine Dykes, Alex Eller, Shayle Kann, Salim Morsy, Flemming Rasmussen, Venkat Srinivasan, Richard Swanson and Ryan Wiser.

10

Manufacturing's new materials

Paul Markillie

A combination of new materials and new techniques will change both what can be produced and where it is made

THE BMW i3 IS A SNAZZY ELECTRIC CAR and, as might be expected, is packed with new technology. Yet the most telling innovations come from the material used to make the vehicle and the way it is constructed. The material is carbon fibre, an extremely strong but lightweight composite. It is turned into a car in a process that is more familiar to the textiles business than metal-bashing. Such a radical change in the way things are made will transform factories around the world making all sorts of goods. It will upend the traditional economics of manufacturing, disrupting long-established trade flows and supply chains.

Knit me a car

How the i3 comes together provides a tantalising glimpse of how different things will be. Instead of starting out as a slab of steel, this car begins life in a Japanese rayon factory as a spool of polyacrylonitrile, a synthetic thermoplastic drawn into a long string, much like fishing line. This is wound onto a spool and shipped to the US, where it is baked into carbonised strands only 7 micrometres (millionths of a metre) in diameter. Some 50,000 of these blackened strands are then spun together to form a thicker yarn and wound onto another spool. These spools are shipped to a factory near Munich, where the yarn is woven into carpet-like sheets on what appears to be a giant knitting machine. When the sheets arrive at BMW's car plant in Leipzig they are cut into shapes and stacked into multiple layers. In an automated process the layers are injected with resin, pressed together and cured to form stiff, lightweight body parts. Finally, the parts are glued together by robots to form the vehicle's body.

The i3 production line is unlike any other car factory. For a start, it is strangely quiet. There are no thundering presses stamping out metal parts or the crackle from showers of bright welding sparks. Nor is there a giant and costly paint shop to clean and apply anti-corrosion treatments to a metal structure (carbon fibre does not rust). Other differences show up in the company's books: overall, production of the i3 uses 50% less energy and 70% less water than a factory would producing such a car with traditional processes and materials.

The Leipzig factory is in the vanguard of new and improved materials making their way out of research laboratories and into production. This materials revolution involves a lot more than just carbon fibre. There are many other sorts of composites, exotic new alloys, specialist coatings, hybrid materials that are part plastic and part metal, organic materials that incorporate biological functions, and "smart" materials that can remember their shape, repair themselves and even self-assemble into components. Moreover, by fiddling with materials at the molecular level it will become increasingly possible to produce bespoke substances with novel properties and to change how materials work, such as the way they respond to light, electricity, water and heat. At the same time, older materials will progressively get better too.

Central to the success of both new and improved materials is the ability to use them at a commercial scale. This process can take years. Carbon fibre, for instance, has been around for several decades, especially in making fighter planes, golf clubs, high-performance mountain bikes and Formula 1 cars. The attraction comes from the material being stronger than steel but at least 50% lighter. That strength arises from the molecular structure of carbon compounds producing strong chemical bonds, much like those in diamonds. By aligning the fibres at different angles, the strength of a component can be reinforced exactly where needed, building in rigidity here and flexibility there.

As experience was gained, carbon fibre came to be used instead of aluminium in commercial aerospace, because lighter planes have lower fuel consumption and produce fewer emissions. Carbon fibre now makes up about half the structure of aircraft such as the Boeing 787 and Airbus A380 and A350. But it comes at a price, largely because production processes have been expensive, slow and labour-intensive. For specialist low-volume products, like fancy mountain bikes and aircraft, it matters less. But carmaking is a high-volume business.

The dark arts

By finding faster, lower-cost ways of using carbon fibre, BMW is pioneering the entry of the material into mass production. Some analysts predict that carbon fibre will become a mainstream production material by the mid-2020s, replacing lots of steel and aluminium. By 2050, when most cars are likely to be electrically powered and many operating autonomously, lightweight carbon fibre will give these vehicles greater range and improved crash resistance.

Plenty of other new materials will by then also be making their mark in cars and other areas of manufacturing. A number of trends are driving the process. The first is a growing understanding of the properties of substances at the smallest scale. Materials scientists have been steadily increasing their knowledge from a century of breakthroughs in physics and chemistry. And researchers now have better instruments, such as electron microscopes, atomic-force microscopes, mass spectrometers and X-ray synchrotrons, to measure and probe materials in much greater detail than ever before.

That detail extends down to the very building blocks of matter. Every material is made up of atoms and how each atom behaves depends upon which chemical element it belongs to. The elements all have distinct chemical properties that rely on the structure of the clouds of electrons that make up the outer layers of their atoms. The way atoms pair off or share their electrons gives structure to molecules, the smallest particle of a chemical element or compound. Being able to engineer materials at the molecular level removes a lot of the guesswork in dealing with new materials.

This is a huge change from the past. When Thomas Edison demonstrated the first proper incandescent light bulb in 1879 he had to rely on trial and error, testing 1,600 different materials, from coconut fibre to a hair from a colleague's beard, before finding a suitable filament to illuminate his invention. Today an inventor might use a supercomputer in the cloud to search for likely candidates, such as new semiconducting materials to make better light-emitting diodes (LEDs), which are far more efficient at turning electricity into light than hot filaments, and which are now replacing old-fashioned light bulbs. LEDs are an invention of materials science, and by 2050 their successors will no longer be discreet lighting fitments, but illuminating films incorporated into the ceiling panels of buildings. Ceiling manufacturers, then, need to think about becoming lighting engineers to avoid being disrupted by lighting companies becoming ceiling producers. Many other industries will similarly face such shifting areas of business.

Accelerating this process will be big data-gathering efforts, such as an open-access enterprise called the Materials Project, based on a cluster of supercomputers at the Lawrence Berkeley National Laboratory in California. This project is compiling the properties of some 100,000 known and predicted compounds to form a sort of "materials genome". This means instead of setting out, as Edison did, to find a substance with the desired properties for a particular job – conductivity, hardness, elasticity, the ability to absorb or repel other compounds and so on – researchers will in future define the properties they want and their computers will provide them with a list of suitable candidates.

Some of the materials already being searched for will provide alternatives to silicon for making faster, more powerful computer chips and to produce better batteries. This may well include graphene, a "wonder material" only one atom thick discovered in 2004 at the University of Manchester. Many other nanomaterials are now being developed. The reason they are of such interest is that unusual phenomena occur when matter is organised at such a tiny level. With modern processing techniques it is possible to turn many bulk materials into nanomaterials in order to take advantage of new or greatly enhanced properties. These include unique physical, chemical, mechanical and optical characteristics, all related to the particles' size.

The breakthroughs to come will change not just products but also people's lives. Better rechargeable batteries will deliver greater range to electric cars of the future and power for longer a host of mobile gadgets, ranging from smartphones to household robots. They also hold out the prospect of recasting the market for intermittent renewable power by storing electricity on the grid, as well as in buildings and homes that generate their own solar and wind power. By 2050 many homes and businesses will find a new level of independence by moving off-grid.

Not all new materials will meet expectations, as many will fall by the wayside as they fail to scale up for commercial use. But just as the process of discovery is being helped by powerful computers, so too is industrialisation. More and more products will start as virtual prototypes in sophisticated three-dimensional computer-aided design and engineering systems, long before anything physical is made. A new car, for instance, can be styled, its engine and suspension tweaked and its aerodynamics refined by a computer. It can also be taken for a test drive using virtual reality through towns and along different roads. These design and engineering systems also take into account the properties of materials, such as loads, stresses and thermodynamics. This makes it easier to explore how new materials can be used to improve a product. And the same computers can be used to design and simulate the production systems required to turn virtual ideas into solid reality.

The printed world

Often that reality will mean not just revamping existing production methods but coming up with entirely new ones. One process attracting a lot of attention is additive manufacturing, or 3D printing as it is popularly known. Although 3D printing began in a basic form in the 1980s, it is only in recent years that improved hardware and software have led to a broad range of 3D printers costing from under $1,000 for hobbyists to over $1 million for specialist engineering applications.

These machines now use dozens of different methods to print objects in materials ranging from plastic to glass, metals, ceramics and even biological substances. Nevertheless, the fundamental principle behind the technology remains the same: building up layers of material additively instead of removing them by cutting, drilling and machining, as is done in traditional manufacturing. There is less waste because a 3D printer deposits material only where it is needed. The machines can also produce complex shapes with geometries that are difficult or impossible to make with conventional production tools – even structures inside an otherwise solid object (because it can be built in layers from the ground up).

At first, 3D printing was used primarily for rapid prototyping, which largely involves making one-off items quickly and cheaply. It can be slow and expensive to set up traditional machines in a factory to make just one thing. But 3D printers are driven by software – the same software that is used to design products – so making one thing and then making another that is different involves only a tweak of software. Additive manufacturing is now steadily producing more things as final products.

Some people have speculated that in the future every home will have a 3D printer making products from software designs downloaded from the internet. For the next half-century that is fantasy, apart from hobbyists and committed do-it-yourself enthusiasts. Nevertheless, 3D printing will become an integrated part of mass manufacturing. Terry Wohlers, an industry consultant, predicts that the market for 3D printing will grow from $6.7 billion in 2016 to $1.13 trillion in 2040 (see Figure 10.1).

Some big manufacturers are already well ahead in additive manufacturing. General Electric (GE), for one, has installed a $50 million 3D printing facility at its factory in Auburn, Alabama, to produce fuel nozzles for its new LEAP jet engine from a "super alloy" made up of cobalt, chrome and molybdenum. A fuel nozzle is a complex component that has to withstand extremely high temperatures and pressure. It is usually made from 20 or so different components welded together. For the LEAP engine, GE is printing the nozzles in one go as a single part by depositing successive layers of the material in a powdered form and melting the required shapes with a computer-controlled laser. The resulting fuel nozzles are 25% lighter and five times more durable than the old sort. GE expects to print 100,000 fuel nozzles a year by 2020.

Airbus has also come up with a proprietary new material, one that it calls Scalmalloy. This is an aluminium-magnesium-scandium alloy which the European aircraft-maker thinks will be particularly good for making lightweight, high-strength metal components in aircraft, particularly with a 3D printer.

FIG 10.1 **Adding up** 3D printing, market value, $ billion

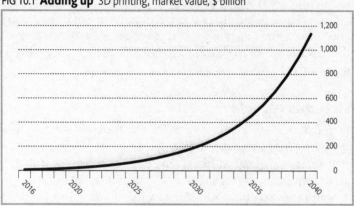

Source: Wohlers Associates, Inc

The first 3D printer has already appeared on a mass-production line in China, and more will follow. LITE-ON, a leading contract manufacturer, uses printers made by Optomec, an Albuquerque-based firm, in its Guangzhou factory to print electronic circuits, such as antennae and sensors, directly into products such as mobile phones and other consumer electronics, instead of making those components separately and assembling them into the devices either by robot or by hand.

The range of things that can be produced with additive manufacturing is growing. At one end of the scale, Winsun, a Chinese firm, is printing houses. The company uses an extrusion head, much like that used to ooze icing onto a cake, to deposit a fast-drying mixture of cement and recycled construction waste to prefabricate large sections of a building which are then joined together on site. A more high-tech approach is being taken by Oak Ridge National Laboratory in Tennessee, which has worked with Skidmore, Owings and Merrill, a firm of architects, to print building structures using materials that integrate components such as insulation, air and moisture barriers and exterior cladding. The idea is to develop an additive building process that results in no waste.

At the other end of the scale, Scrona, a company spun out from ETH Zurich, a Swiss technical university, is 3D printing some of the smallest objects. It uses a process called NanoDrip which, as its name suggests, deposits tiny droplets of liquid containing nanoparticles – measured as 100 nanometres (billionths of a metre) or less – to build microscopic structures. These include silver or gold conductive grids, invisible to the naked eye, which can make touchscreens more responsive to finger movements.

Virtuous recycle

Nanotechnology will provide ways to enhance the abilities of materials which have long been used in manufacturing. Modumetal, a Seattle company, is using nanomaterials to build up veneers of different metals to produce what it calls "nanolaminates" by means of a type of electrolytic deposition, which is a bit like electroplating but more sophisticated. By carefully manipulating an electric field the process gets various metals suspended in a liquid to form layers on the surface of objects. The technique also controls how the materials in the nanolaminate react with each other. The company has begun coating components used in the gas and oil industry to provide what it claims will be an anti-corrosion protection that will last up to eight times longer than conventional treatments.

In the future, Modumetal believes it will be possible not only to coat structures but also to "grow" complete components from nanolaminates using conventional materials such as steel, zinc and aluminium. Moreover, it is possible for the electrolytic process to be reversed. This means once nanolaminated components reach the end of their working lives the material used to produce them can be recovered.

As materials become ever more exotic, recycling will become a necessity in manufacturing. New ways will be needed cheaply to dismantle products, such as electronics, and recover the materials they contain. Cars made out of steel and aircraft produced with aluminium are relatively straightforward to recycle, but this will be more difficult as increasing amounts of carbon fibre are used to reduce weight in the transport industries. Companies are working on ways to recycle carbon fibre, in some cases chopping it up and using it to produce lower-grade components, such as panels that are not subjected to large amounts of stress. A growing number of rechargeable batteries will, in time, also need to be recycled. Some nanoparticles are known to be toxic, but scientists do not yet fully know what their long-term effects will be if they end up dispersed in the environment. Nanoparticles are being deposited on land and washed into rivers and the sea. Tonnes of nano titanium dioxide, used in sunscreen lotions, are washed down drains every year.

Manufacturers will be made increasingly responsible for the so-called life cycle of their products, from the extraction of the elements used to make their materials to where these materials eventually end up. This will also be driven by commercial concerns: some of the elements used in advanced manufacturing are rare and expensive. Urban mining will become a big business, retrieving traces of materials such as gold, silver, neodymium, yttrium and dysprosium from dismantled gadgets, electric cars, batteries and household goods. Computers, once again, will help companies by modelling product life cycles and simulating how products can be dismantled and recycled.

All these things will transform the way manufacturing has worked over the past century. For most of that time making things has been largely a "me too" game, with factories using the same basic materials and employing similar production methods and equipment. Such things are easily copied, so for mass-produced goods it meant that economies of scale and wage costs mattered most, driving much manufacturing offshore to low-cost countries. By 2050 a lot of production that went offshore will have moved back.

Manufacturing's homecoming

The tide will turn for a number of reasons. For a start, new materials and novel production technologies allow greater manufacturing flexibility, especially in the customisation of products and by making goods more efficiently in lower volumes. This means it will pay for companies to locate closer to the markets they sell into, so products can be tailored more specifically to those markets and to respond to changing trends much faster.

Wage costs will still matter, but less so with automation taking over most monotonous jobs. Many items will still be made overseas, but more in specialised clusters of excellence – like that for electronics in southern China – rather than low-wage sweatshops. This will be particularly so with commodity-type components, like computer chips and other electronic widgets. However, the factories where such components get incorporated into finished products will become more widely dispersed and, importantly, bespoke.

Companies will increasingly have unique production processes, often tailor-made for their needs and materials. Just as BMW has come up with its own particular way to make electric cars, and GE has developed proprietary materials and methods to print its fuel nozzles, others will follow. Nike, for example, has also turned to knitting. It has a process called Flyknit, which can produce a pair of trainers on computerised knitting machines using a special micro-engineered yarn, instead of having to stitch shoes together by hand from cut-out panels, a job largely offshored to factories in Asia. Nike's automated knitting machines could be operated anywhere, even inside high-street stores, making customised trainers to order from scanning customers' feet. Adidas, a Nike rival, is bringing some production of its trainers back to Germany with a new highly automated factory near Ansbach.

Bespoke materials and processes are much harder for rival firms to imitate, so they will become a source of competitive advantage. Often it is not the technology itself that represents a company's "secret sauce", but the carefully developed processes wrapped around it and the talent employed to operate it. And that provides another factor contributing to where production will locate: wherever people with the required skills, particularly in design, materials science, software and engineering, can be recruited. Such people will be a valuable asset and in great demand to run the factories of the future. The days of dark satanic mills, oily rags and overalls will have gone. The great entrepreneurs of the future will be making things.

11

Military technology: wizardry and asymmetry

Benjamin Sutherland

The West will benefit from big advances in weapons and information technologies, but these will also create new vulnerabilities that foes will seek to exploit

THE DISTANCE FROM WHICH WESTERN SNIPERS regularly score kills has nearly doubled in a mere two decades, notes Tom Gil, who, as an Israel Defence Forces sharpshooter, not long ago shot a man dead at more than 1,800 metres (1,970 yards). Rifle systems have become so good that kills from even farther away are no longer uncommon. Breaking a combat record in Afghanistan in 2009, a British sniper, Craig Harrison, felled two Taliban machine-gunners from a dizzying 2,475 metres. The bullets had flown for nearly six seconds.

Stunning advances of this sort have been achieved in a wide range of military technologies in recent decades, and yet much of this progress will be dwarfed by wizardry likely to emerge in the coming decades. Some of this will most benefit the West, partly (but not only) thanks to superior defence R&D. Other developments, however, will threaten the most technologically advanced forces of the US and its allies, limiting the areas in which they can operate with a degree of safety from rival forces such as those of Iran and "near peers" China and Russia. This poses a big challenge to the West. In 2014 Chuck Hagel, then US secretary of defence, warned: "We are entering an era where American dominance on the seas, in the skies and in space can no longer be taken for granted." Add cyberspace to the list, says Kenneth Geers, a NATO adviser studying the "mind-blowing progress" in state and non-state cyber capabilities in Russia and Ukraine from his base in Kiev.

The free world has twice overcome strategic threats to its military dominance. After the Soviet Union and China first detonated nukes, in 1949 and 1964 respectively, the US and allies in western Europe fostered market economies vibrant enough for an edge in the engineering and industrial base needed for superior conventional forces. By the time the Russians began catching up in the late 1960s, the West's second successful "offset" was under way in the form of computing. This led to better spy satellites and guided bombs and missiles, demonstrated to devastating effect in the first Gulf war in 1991. But as computing and satellite technology spreads, the US's lead looks vulnerable once again. Hence the quest for a "third offset", again based on Western technological prowess. Success is by no means assured.

Creative destruction

By mid-century sniping will have made further strides: elite forces will probably shoot guided bullets. The US Defence Department has already begun work on a bullet called EXACTO with fins that adjust its trajectory in mid-air. Freed of the need to be in the target's line of sight, snipers will be able to devise devastatingly imaginative shots that skirt obstacles. To show the bullet's optical system what it needs to hit, a laser aiming an invisible infrared beam will need to be able to pinpoint the target, but this kit will be mounted on imperceptible drones high above.

Guided bullets promise plenty of misery for combatants who don't wield them. The technology will dramatically lengthen elite snipers' range as the extra distance provides a bullet with more time to correct for faulty aim or unanticipated wind. Imagine the damage to an enemy's morale when it begins losing fighters to "an invisible shooter" so far off he could not be reached with return fire even if he could be spotted, says Ryan Innis, until recently a sniper with a US Marines unit combating piracy off east Africa.

The role of sniper fire, and therefore guided bullets, will expand in coming decades. The growing capabilities of drones, aerostats and satellites to spy and guide missiles will make it harder for irregular forces to find sanctuary in mountainous terrain. This will shunt more guerrillas into cities. Guerrillas in urban areas can hope to enjoy some protection from Western forces, whose aversion to civilian casualties will continue to grow in a digital age that facilitates the use of violent imagery for powerful propaganda. Sniping causes less collateral damage than artillery, bombing and strafing, so Western forces will use it more.

As 2050 approaches, infantry from advanced forces will also become harder to kill. The ballistic vests and helmets worn today cover just 19% of the body, for more armour would be impractically heavy. But as gear lightens in the years ahead, elite soldiers will don more armour. Efforts under way to make bullet cases with polymers rather than brass should cut the weight of a soldier's ammo by a third. Moreover, armour itself will become lighter. Engineers at Moratex, a research institute in Lodz, Poland, are developing a "non-Newtonian fluid" that becomes viscous enough when struck to stop a bullet. These "shear-thickening" fluids are lighter and more flexible than the Kevlar and ceramic plates used today.

Some elite Western soldiers will be fully encased in powered, bullet-proof exoskeletons. One such "Iron Man" suit being prototyped by the US Special Operations Command, dubbed TALOS, will eventually integrate weaponry, monitor a soldier's condition and give him superhuman strength. Those who wish to kill Western soldiers will no doubt also be frustrated as robots increasingly take their place in the air, on land and at sea.

Military robotics are already so good that Boeing and Northrop Grumman, two US defence firms, are building unmanned fighter jets, respectively the X-45 and X-47. With no need for crew facilities, these drones will provide more payload, range and stealth for the buck. In 2015 the US Navy secretary, Ray Mabus, said that Lockheed Martin's F-35 will almost certainly be the last manned strike aircraft it will ever buy. By 2050 drones will range from stealthy spies resembling flying insects to doglike resupply and attack models that can live off the land for months at a stretch, combusting leaves and wood for power.

Unmanned and especially autonomous systems must become, Mabus says, "the new normal". Just how independent drones should become, however, is controversial. Defence chiefs say decisions to fire will be made by "a man in the loop", not robots. But exceptions are already made, for example with robots that defend warships from missiles approaching too fast for sailors to react. And, crucially, technologists are designing software that lays the groundwork for greater decision-making by drones.

The US Pentagon funds the development of "ethics" software. These programs will query databases to determine if, say, a missile fired from a certain position might violate norms of just war, perhaps by spraying shrapnel into a schoolyard on a school day. The goal is to aid, not replace, human decision-making. But, troublingly, China and Russia are believed to be designing such software with an eye to removing the human from some firing decisions. Responsibility for deaths is thus destined to become "very, very, very diluted", says Emmanuel Goffi, a French air-force expert on robotic warfare. That's a problem, he says: you can't pin war crimes on a robot.

Warfare's robotised future will bring other headaches. As robots replace soldiers, some armed groups, frustrated by the difficulty of getting an enemy human soldier in their sights, may increasingly target civilians instead. And the growing capability of robots may encourage countries to begin unwise wars, if politicians reckon that an attack that can be launched without putting soldiers on hostile soil carries seductively little political risk.

Another matter is how long the West will hold its lead in military robotics. Rivals are advancing. In 2015 Russia's deputy prime minister, Dmitry Rogozin, announced that UralVagonZavod, a military manufacturer, would turn its T-90 tanks into robots remotely operated by soldiers with the skills of video-gamers. Earlier, he had written of a future in which a Russian "army manned with bespectacled nerds would completely destroy the forces of handsome athletes who fight on a lower technological level".

Fire and brimstone, exquisitely engineered

The robotics technologies that most trouble Western strategists are those which, coupled with rocket science, will lead to rivals' development, or import, of devastatingly precise missiles, large and small. As China's State Council Information Office wrote in a 2015 strategy paper, the army's accelerating "evolution to informationisation" would beget increasingly "precise, smart, stealthy" missiles able to hit distant targets, most importantly at sea.

The forces that will wield such munitions for the first time in coming decades include non-state groups. Those without a spacefaring benefactor to provide satellite reconnaissance for the monitoring of an enemy and selection of target co-ordinates will simply purchase the data, perhaps through a third party for greater discretion. A firm near Tel Aviv, ImageSat International, offers "rush satellite tasking" as part of "intelligence as a service".

By mid-century the West's already slipping monopoly on precision warfare may well be long gone. Can Western powers concoct kit to compensate for such a momentous shift, thereby continuing to protect their tanks, aircraft and warships?

"Electric armour" may help. The UK defence ministry's Defence Science and Technology Laboratory hopes that dumping electricity from a capacitor into armour a moment before a warhead hits will sharply limit damage. Work on exquisite but expensive systems to track and shoot down incoming missiles will also continue. Even so, the spread of terrifyingly accurate and destructive guided missiles promises disproportionate benefit for today's underdogs.

The stakes are highest for the US Navy and the liberal world order that depends on it remaining top dog. China, Russia and others will design and export better missiles. Iran can expect particularly big advances in missile prowess – a future boon, no doubt, to its Islamist clients, including Hizbullah, a Lebanese militia, and Houthi rebels in Yemen. As part of the 2015 nuclear deal, Western powers, under pressure from Russia, agreed to lift embargoes on Iran's import and export of technologies for conventional arms and ballistic missiles in 2020 and 2023, respectively.

Two of today's guided missiles provide an inkling of future capabilities. Russia's Kalibr, dubbed the Sizzler by NATO, "sea skims" for hundreds of kilometres, skirts obstacles, dodges countermeasures and sprints at three times the speed of sound to better punch 450 kilos of explosives into a ship's hull before detonating. Export versions have gone to Algeria, China, India, Vietnam and, some suspect, Iran. A four-missile Kalibr launch package disguised as an ordinary shipping container is marketed for discreet emplacement on a commercial ship, train or lorry. The DF-21D is a "carrier killer" ballistic missile made by China. Shown in its first parade in Beijing in 2015, it sports a manoeuvrable warhead packed with enough explosives to devastate a large warship even 1,500 kilometres (930 miles) away.

Such missiles have led the US to "spend rather wildly" on fancy kit to defend ships, says Jerry Hendrix, a former head of the US Navy secretary's advisory panel. By mid-century US laser guns will shoot down missiles even 10 kilometres away, reckons David DeYoung, head of laser weaponry at Boeing. A Boeing laser demonstrator sold to the US Army already destroys incoming mortar shells at much shorter range, he says, noting that fear of missiles like the DF-21D is driving R&D in "directed energy", jargon for laser weapons. Missiles might also be shot down with slugs hurled more than five times the speed of sound from big railguns fed a massive jolt of electrical energy.

But the effectiveness of those countermeasures will be iffy, says Hendrix, whose responsibilities included forecasting capabilities. The smarter, longer-range missiles of the future promise to turn much of the ocean into a maritime no man's land. The US should therefore stop building $13 billion aircraft carriers that will be threatened by missiles with much longer range than the air wing tasked with knocking out launch facilities, he says.

Premvir Das, a former commander of India's Eastern Naval Command, says that "the whole envelope of operations at sea" will change. Submarines, and especially nuclear ones, will take on greater importance, Das says. Surface vessels' growing vulnerability will push more naval activity underwater, with submarines expanding their role to include stealthily moving into position elite troops as well as spy and attack drones that operate under water, on land and in the air.

The best submarines cannot be made much quieter than they already are, however. So, as sensors improve, they will become easier to locate and therefore sink, says Andrew Krepinevich, a former adviser to three US secretaries of defence. The advantage conferred by taking the first shot will therefore grow, reckons Krepinevich, now head of the Centre for Strategic and Budgetary Assessments, a think-tank in Washington, DC. This is troubling. Amid rising tensions, powers at a technological disadvantage will have extra motivation to be the first to launch torpedoes or missiles – potentially starting combat that might otherwise have been avoided.

If the liberal world order is to prevail, the West must, somehow, compensate for the spread of fearsome precision munitions among the authoritarian and extremist forces challenging it. The best hope for a "third offset" lies largely with artificial intelligence, believes Krepinevich, who is now on the Pentagon's Defence Policy Board. It will provide, in addition to autonomous robots, clever ways to exploit "big data" militarily. To illustrate this, he points to the US Navy's processing of undersea acoustic signals during the cold war. It was so slow that a Soviet submarine would often be long gone before it was noticed. Future algorithmic wizardry may sufficiently speed up the recognition of patterns in reams of data from undersea sensors for the US to pinpoint enemy submarines in real time at great distance, he says.

But the West is likely to find it ever harder to hang on to military advantages derived essentially from superior computing. Innovations devised in Western universities spread quickly. Computing breakthroughs will power products sold by the likes of Apple and Intel to anyone. And because consumer and industrial spending dwarfs that for stuff with exclusively military applications, technological advances for specific defence purposes will increasingly lag behind those built into components or systems available to all. Boeing bought the laser it put in the US Army's anti-mortar system as a "commercial off-the-shelf" product.

Battle of ideas

The West owes its edge to more than the stuff it makes, however. Soldiers raised in a Western liberal democracy also have a cultural advantage in combat with enemy fighters who grew up under a theocracy or any kind of authoritarian regime. With a background of greater empowerment to make important life decisions and recover from mistakes, Westerners tend to display greater creativity and innovative spunk. This translates into a greater ability to modify tactics in battle to better seize unexpected opportunities that soldiers from undemocratic countries may not have the reflexes, or even permission, to exploit.

The designers of software that forecasts the outcome of a hypothetical or likely future battle, campaign or war refer to this cultural advantage as "initiative". By studying past combat between forces that have initiative (such as Israeli troops) and those who mostly do not, forecasters have devised algorithms that quantify the advantage this inventiveness is likely to provide in numerous scenarios. Often, it's a lot.

This edge will grow along with advances in robotics and communications kit. The reason is that these technologies will increase the amount and quality of tactical intelligence gathered and delivered to troops, and Westerners, crucially, have a cultural advantage in its use. Education systems in democracies encourage the type of creative problem-solving that is facilitated by timely intelligence, be it the likelihood that agricultural run-off following rain has polluted a well, say, or an analysis of a town leader's motives for whipping up anger at the government.

It helps that Westerners are used to processing information that is neither controlled nor distorted by an illiberal regime. Autocrats' restriction of freedoms to tighten their grip on power leaves them with soldiers disadvantaged by a relative lack of critical thinking and associated ingenuity, says Pieter Cobelens, a former director of the Netherlands' Defence Intelligence and Security Service. Their handicap will only grow, he adds, as new technologies make battlefield success less a matter of blind obedience and more a result of the clever use of detailed intelligence and precision weaponry.

The success or failure of a third offset will depend a lot on how much the West squeezes from its broad cultural advantage. To make the most of it, the US will craft algorithms that determine which soldiers should receive which bits of intelligence, says David Shedd, until recently acting director of the Pentagon's Defence Intelligence Agency. Automating the "pushing" of intelligence to soldiers in this way will free them from the need to surmise what might be helpful and then dig it out of vast databases. Exquisite visualisation technologies, he adds, will be developed for this. Before 2050 Western soldiers will assimilate tactical intelligence without taking their eyes off their surroundings. The screenless displays necessary for this are already in the works.

By mid-century, some reckon, headset-mounted lasers will scan imagery directly onto soldiers' retinas. Others, including a US firm, Avegant, see more promise in using silicon chips bristling with tiny hinged mirrors that flip back and forth to reflect bits of multicoloured LED light onto a viewer's retinas. It sells a headset named Glyph that uses two chips with more than 1.8 million mirrors, each 5 microns across, which flip at least 3,600 times a second to generate an image that appears to float in mid-air without blocking out the surrounding world. The firm's co-founder, Edward Tang, foresees an eventual "augmented reality" capability that superimposes tactical intelligence on relevant objects, as the beholder shifts gaze.

These advances will not be trivial, says James Geurts, procurement boss for the US Special Operations Command. By displaying intelligence on the objects it relates to, the technology will deliver, he says, "meaningful, tactically relevant information at the point of need – that's the Holy Grail". Markings that indicate suspected insurgents' hideouts, say, or the locations of previous improvised-explosive-device blasts will appear to hover over the right spots, even as soldiers walk and turn their heads. The idea, he says, is to better harness American soldiers' competitive advantage of "adaptive velocity" in the face of rapidly changing circumstances.

Supremacy of the heavens

Any primacy that relies on data processing will require satellites to collect and transmit information. But China and Russia, like the US, can already strike satellites and before 2050 at least 16 more countries will acquire "anti-sat" capabilities, says Erwin Duhamel, an official in Belgium's defence ministry who was previously the country's top military co-ordinator for space. The list will include, he believes, Brazil, India, Iran, Nigeria, Pakistan, South Africa, Turkey, Vietnam and, if the regime survives, North Korea.

Today's big lorry-sized satellites will therefore be replaced by "constellations" of smaller networked modules that will be more difficult and expensive to knock out. So reckons Dennis Göge of the German Aerospace Centre, a government space-robotics lab in Cologne. A few powers, the US and France included, will develop an ability quickly to build and launch small replacement satellites for emergencies. Studies conducted by Dassault, a defence company, suggest that France hopes eventually to be able to launch, in a pinch, small satellites from a modified fighter jet based on an aircraft carrier.

Even at mid-century, however, satellites will remain easier to destroy than to defend or replace, especially if debris from strikes starts a chain reaction, Göge says. He believes that everyone's interest in limiting debris in low-Earth orbit will prevent attacks. Others are less optimistic. Will a regime in a bruising war, perhaps facing collapse, show deference for the global commons?

George Friedman, founder of Stratfor, a geopolitics consultancy in Texas, thinks that the US will arm its satellites with defensive missiles to smash incoming interceptors and anti-sat laser guns. Over time, these satellites, he believes, will evolve into "Battle Stars" with offensive arsenals to wage war from space. Holding the "high ground" beyond many adversaries' reach, Battle Stars would, in theory, give the US and close allies a big new edge.

Some military thinkers, however, see future wars moving more into cyberspace than outer space. However fearsome, Battle Stars will not be worth much to a country that loses control of its computer networks. Cyber-firepower could become more determinative than the kinetic sort, says Cobelens. "Who cares about your airplanes if your air traffic is not OK, if your electrical power can be cut off, and they erase your name and let your money vanish?" he asks.

Such a prospect certainly sounds frightening, partly because a future world vulnerable to "logic bomb" attacks will offer an asymmetric advantage to non-state groups hard to hold accountable or even identify. But might cyber-weaponry usher in a future in which conflict is less destructive?

Recent technological progress has tended to make warfare less bloody. Because precision strikes can destroy critical enemy systems, or the specialists needed to operate them, there is less need to chew up masses of infantry. In 2003 US-led forces rendered the 380,000-strong Iraqi army mostly inoperative by killing fewer than 11,000 troops in just three weeks. In contrast, the lower-tech 1980s Iran–Iraq war dragged on for eight years despite deaths numbering at least 250,000 for Iraq and perhaps 1 million for Iran. Cyber-weaponry could conceivably nudge this trend along. Why kill people if disabling their computers achieves the same military goal?

The development of digital weaponry will accelerate. David Lindahl, an expert on cyberwar at the Swedish Defence Research Agency, a defence-ministry body in Stockholm, believes secret efforts have already begun to create digital viruses that will be injected into computers through thin air via exquisitely crafted blasts of radio waves or microwaves. For now, though, no one knows if computer worms of the future will provide a viable way to wage war, he says. The Stuxnet virus discovered in 2010 which had been developed, probably by the US and Israel, to stop Iran's effort to enrich uranium, he notes, succeeded only in "hampering it somewhat".

Known unknowns

By 2050, in any case, stunning advances in military kit and capacity may be only part of the story. Martin van Creveld, an Israeli military historian, worries about another side. A lot of the West's elaborate military R&D, he says, increasingly serves as a dangerously comforting substitute for a declining will to fight, epitomised by timid resistance to Islamic State in Iraq, Syria and Libya. This has happened before. The decadent, declining Roman Empire put more effort into engineering better catapults than actually fighting barbarians, van Creveld says.

As worrisome, some reckon, is that some countries' advances in military technologies will encourage adversaries that feel they cannot keep up to compensate with nuclear weapons. Tellingly, Pakistan and Russia, both faced with the technologically superior rival forces of, respectively, India and NATO, decline to forswear the first use of nuclear weapons. Further, the inevitable spread of expertise to build and deliver small "battlefield" nukes will increase temptations to use nuclear weapons. This is one more possible ramification of advancing military technologies that strategists would be well advised to keep in mind.

12

Personal technology gets truly personal

Leo Mirani

As the real and virtual worlds become increasingly intertwined, digital technology will penetrate ever deeper into our lives, and quite possibly into our bodies

IT WAS ONLY 20 YEARS AGO that the default mode for even the most enthusiastic geek was offline. Connecting to the internet meant going to the internet, like fetching water from the village well. To get online, early users would dial a phone number using a modem that sat beside a towering desktop computer, waiting patiently as the connection bleeped and blooped to life. Today, in the rich world, the internet is all around us, with ubiquitous Wi-Fi and mobile coverage. The transition from a default offline world to a default online one is largely not remarked upon today. Like running water, it is something noticed only by its absence.

Something similar will happen over the next few decades to the devices that serve as our gateways to the internet. They will disappear. Sitting at a desk with a computer or whipping out a smartphone will seem as old-fashioned as manually making a connection to the internet. Indeed, the word computer itself will drop away from the lexicon. When everything around you is capable of computation, what constitutes a computer any more?

This change will be driven by huge advances in virtual reality, its cousin augmented reality, and a family of related technologies that will allow us to interact with the cloud (or whatever we are calling it by then: "the world", perhaps?). It will drive a shift in human behaviour bigger even than that caused by the advent of smartphones and the web. Personal technology will at last be truly personal.

VRooom

Start with virtual reality (VR). The most remarkable thing about it – the thing that makes it unique in the rich history of human communication – is "presence". This is the deep, visceral feeling of "being there". Those who have experienced it recount memories from VR as though it were a place they visited, not something they saw on a screen, even if the experience itself is of a low-fi, obviously computer-generated world. In 2017 that still inspires a feeling of awe. By 2050 it will be the most natural thing in the world.

What are the benefits of being there? The first application is entertainment. It will not take until 2050 for people in the Western world to start attending concerts and sporting events from the comfort of their couches. Nor need it cost the earth – just as 3D movies are slightly more expensive than their 2D versions, so VR entertainment will be priced comfortably within reach of those who already pay big sums for pay-per-view sports and other events. As more people join in, the prices of the devices will drop.

As the cost declines, so too will the heft of the gadgets. The first VR rigs, invented by VPL Research, a pioneer in the field, were huge things, bulky body suits with lots of trailing wires, data gloves and heavy headsets that looked as if a mechanical octopus had attached itself to the top of the wearer. (A later, smaller version cost some $9,000 and was called the EyePhone.) Today they come in two forms. The simpler version is a cradle to slip a smartphone into, such as Samsung's Gear VR or Google's Cardboard (literally a piece of cardboard). The other type is exemplified by the Playstation VR, Oculus Rift or HTC Vive, which come with built-in displays but rely on external processing power (typically in the form of a computer or game console). It is a safe bet that by 2050 even the most demanding devices will not require external processors, and that they will be lighter and smaller than anything that exists today.

The second early application of virtual reality is gaming. Gamers have always demanded faster processors, better screens and more reliable connections, happily paying large sums for the privilege of being at the cutting edge of technology. They will do the same again with VR, creating the initial market for new products and giving manufacturers a sample group on which to test new ideas. In the 33 years since Tetris was first released, computer games have become eerily lifelike and immensely complicated, with computer graphics that rival anything in big-budget superhero action movies. The year 2050 is another 33 years away – and the rate of improvement in computer graphics is only accelerating.

These early, easy wins will lead to more useful ones: doctors examining patients from afar; immunocompromised children attending school without fear of catching bugs; factory inspectors checking products via remote robot; soldiers training for unfamiliar terrain; business negotiations where participants can see every fidget of their counterparts – the list goes on.

Yet it is when you look beyond the obvious that VR becomes truly compelling. Chalktalk, a program created by Ken Perlin of New York University, suggests one possible future. Chalktalk is a virtual pad on which its users can sketch anything – shapes, graphs, computer code, mathematical equations – just as they would on a blackboard. The difference is that the shapes become three-dimensional objects, the equations work, the code compiles. In one example, Perlin draws a pendulum and sets it swinging. The swings are measured on a graph, which is also drawn. In another, he makes a chart, which resolves into a 3D graph. A matrix of logarithms influences the curves. In a third, he sketches a vase, finessing it and refining it until it becomes a fully formed 3D object. Step out for a coffee just a decade or two in the future, and the 3D printer in the corner will have finished manufacturing that object by the time you are back. These far-out ideas are already possible using computer screens and existing technology. But it is easy to see how advanced versions of similar ideas might be used in virtual reality for teaching, collaboration, business, or applications yet to be imagined.

An augmented world

If presence is what makes VR so uniquely powerful, it is also what limits it. This is where augmented reality (AR) comes in. If virtual reality is something that requires being in a delineated space where you don't crash against walls or coffee tables, augmented reality is made for the outside world. AR is to smartphones what VR is to desktop computers.

Again, existing technology hints at what is possible. Pilots have for years used "heads-up displays" projected onto the front of their cockpits or on visors mounted on their helmets. These displays are becoming common on car windscreens too. But this is the most primitive form of AR. A slightly more advanced version is Google Glass, which displays information in a pair of spectacles worn like normal glasses. But it shows only a small rectangular display, which is not that much better than a screen held in a hand and seen at a distance. Magic Leap, a secretive Florida-based start-up, goes one better: its technology shows 3D objects that bear some relation to the objects around them. Yet these too are for the moment novelties – models of the solar system, for example, rather than useful information overlaid onto the real world. Indeed, that is what makes AR trickier than VR: the glasses must not only display information but also map and understand the physical world, sense depth and distance, crunch data to figure out what they are looking at and place objects in the correct positions.

By 2050 this will be the norm. In advanced societies, AR glasses will have replaced smartphones for all but the most technophobic. No longer will directions be displayed as blue lines on flat smartphone screens. Instead they will appear as trails to be followed on the streets ahead. Restaurant menus will be redundant. Walk past a café and its entire selection will be available to scroll through, with steaming projections of what the dishes look like. Conversations with people who speak other languages will be simultaneously translated. Plumbers will go out of business as detailed visual instructions for fixing a blocked sink will be readily available, overlaid on the problematic drain. Buses need not display information. Your glasses can tell you the bus number, its destination, its route and the expected time of arrival in a language of your choice. You will never forget a name again, since everything you know about a person will flash by their faces as you converse. Shops need have no signage. Municipal authorities could scrub away the road markings and signs that mar our cities.

The visual clutter of early 21st-century life will be replaced by pristine environments in which what we see depends only on what we need to know, and nothing more. We will also be able to decide what level of reality we want. Most of it? Or as little as possible? We could spend our days wandering around 14th-century versions of our cities if we so desired, and still be fully functioning creatures of the 21st. Just as no two smartphones are the same once you turn them on – each user has a different set of apps, shortcuts and contacts – so will the world appear different to each one of us.

If this sounds far-fetched, consider that many newspapers no longer publish print editions, that London buses no longer accept cash as a form of payment, that there are taxis on our streets that bear no external signage and can be hailed only through an app (and for all practical purposes are invisible as taxis to those without smartphones), and that all of this has happened within a decade of the introduction of the first iPhone.

Why this time is different

The coming of virtual reality has been predicted for at least a quarter of a century. The latest wave of excitement – triggered in large part by the purchase of Oculus, a VR start-up, by Facebook for $2 billion in 2014 – brings to mind the optimism of the early 1990s. But there is cause to believe that this time is different.

First, the number of people who use computers is manyfold larger. The pool of early adopters who might cough up cash to buy the latest thing is correspondingly much larger as well. Second, the amount of money needed is tiny by most standards. In 1990 a prototype VR headset cost nearly $10,000; in 2016 the Oculus Rift was priced at $599. In less than a decade the price is likely to have dropped by an order of magnitude. By 2050 it will be cheap enough to reach much of the world, not just the rich bits. Third, VR has gone from something that was the domain of geeks in Silicon Valley to a technology for which every major entertainment company is preparing content. More and more film festivals have a VR section; gamesmakers are releasing titles in VR. In the early 1990s, studio executives would say, of embracing new technologies, "Why bother?" recalls Nick Demartino, who at the time ran a technology studio at the American Film Institute. Today, he says, they are terrified of missing out.

The fourth reason for optimism is that the technology VR needs has now advanced to a point where it is not wildly impractical to think that it is poised to take off. The internet is everywhere, processing power is cheap and plentiful, and high-definition displays have been here for years.

But for VR to fulfil its potential, technology needs to progress farther, making both incremental improvements as well as great leaps.

Take incremental improvements first. Already telecoms operators are racing to be the first to offer 5G mobile networks. The vast majority of the world uses either 3G mobile broadband or 4G/LTE, which transmits data ten times faster than its predecessor (the G stands for "generation"). The next version, 5G, will be anywhere between ten and 100 times faster than 4G/LTE, but it will come with other improvements as well, including support for lots of devices at the same time and extremely low latency (the time lost in the process of transmission). High-speed connections are important not only to access information, but also to connect to processing power. As Moore's law, the 1965 rule of thumb whereby processors get twice as fast every 12 months (later amended to 24 months), stops working its magic (see Chapter 4), computation will move to the cloud. For lightweight AR glasses to become a reality, they will need to be in constant communication with bigger computers somewhere far away.

Other technologies will need to improve too. Displays will become lighter, pixels smaller and denser, and computer graphics will have ever more polygons. These things are a matter of when, not if. They are already in development.

Then there are the technologies that are yet to mature. A modern smartphone conceals within it a dozen or so sensors. But that number will explode, both within and outside our machines. The world will need to be crammed full of tiny sensors for our new devices to know where they are and what they are looking at, and to understand space and depth. This is easier to imagine indoors. Living rooms or offices can be kitted out with sensors and 3D projectors. Discreet devices can project lifelike objects or avatars of people, while sensors track our movements and interactions with them. This too is not far in the future: Microsoft's Kinect, a gaming device, can sense movement.

Advances in artificial intelligence and machine learning will also be important. One person using an AR device is a lot less useful than a hundred million people using AR: the usage patterns and behaviours from a massive number of users can be analysed to improve the technology, allowing machines to figure out what people expect when they look at something or tilt their heads in a particular fashion.

Great leaps forward

Farther into the future, it is the great leaps that will make VR and AR appear seamless in 2050. We may experiment with things like sensor-laden wristbands or clothing with circuitry woven into them. As time passes, technology will come ever closer to our bodies – eventually finding its way inside us. It starts with contact lenses instead of glasses. The technology for this, in a rudimentary form, is coming into view. In 2016 Samsung applied for a patent for smart contact lenses.

From contact lenses it is a short leap to imagine a simple operation to replace the lens of the eye with a technologically superior version, perhaps done at birth. While we are getting speculative, why not replace the entire eyeball with one that comprises all the gadgetry required to make AR work? Indeed, as human beings become more comfortable with the idea of implants, technology will burrow its way deeper into us, perhaps concluding with implants in the brain.

That is how we receive information. But how do we transmit it? When *Minority Report*, a film directed by Steven Spielberg and based on a short story by Philip K. Dick, was released in 2002, its vision of the future was one where computers were panels of glass that required touch and gesture to manipulate. This has come to pass; moving our fingers around glass screens has become a natural form of input. Even infants can figure it out. Dale Herigstad, an "advanced interaction consultant" and one of the people who worked on the film's futuristic interface, thinks he didn't go far enough. Why do you need big screens when empty space is a good enough canvas to project upon and draw shapes on?

Just as glass screens as displays will seem archaic by 2050, so too will the idea of bashing away at a keyboard seem antediluvian. Google is working on something called Project Soli, which uses radar to sense the movements of fingers; the idea is that natural actions, such as turning a radio knob or pressing a button, can be simulated without the need for actual knobs or buttons. Herigstad thinks that we will create something akin to sign language: entire new grammars and vocabularies to communicate with our machines, languages that will feel as natural as swiping left on a smartphone screen (which itself is something that did not exist a decade ago).

Or perhaps systems will evolve by 2050 that allow machines to look directly into our brainwaves. This is not as remote as it sounds; at least one company, called Emotiv, has been formed to explore the possibilities of what it calls "brain–computer interfaces". Long before that we will be able to control things simply by looking at them and blinking. The technology for this, too, is in development – and it works, if in a rudimentary fashion.

There is a third, equally important aspect to virtual reality, beyond display and input: haptic (or physical) feedback. Touching a smartphone screen is a satisfying experience because of the resistance offered by a stiff piece of glass. Drawing shapes in the air may work because of visual cues. But what of things that require physical sensation, such as shaking hands? A decade or two ago the answer might have been tech-enabled gloves. But the future offers better. Nonny de la Peña, a pioneer of VR, believes it is sound you cannot hear that carries the answer to haptic feedback.

Sound travels in waves, and as anyone who has ever attended a rock concert can attest, it is possible to feel the bass pulsating through the crowd. At the right frequency, pitched at the correct angle, sound could also provide the sensation of touching something, of shaking hands with a virtual friend several thousand miles away.

None of these systems needs to work perfectly on its own. Nor will they be right for every situation. But working together, along with technologies as yet undreamed of, they provide the basis for a world in which computers have ceased to exist as things we carry around. Instead, they will be everywhere, including inside us.

No hiding place

Society will have to accept certain trade-offs to enjoy this future. The first is constant, near-perfect surveillance. Today, the data collected by your smartphone already know more about you than your partner or your mother. With GPS tracking, motion sensors and call logs, it is possible to draw a rich picture of your day-to-day activities. Add in browsing on social networks and search history, and your smartphone may know you better than you know yourself.

But there are still things that machines don't know about us. With VR, this will change. The companies that make these devices will transmit back to their headquarters every twist of your neck, every flick of your pupils, every reaction to stimulus. This surveillance will reach further in AR: devicemakers will be able to see everything you see. They will, literally, be able to see the world through your eyes.

The companies in charge will claim they have no choice, as these data form the basis on which the service works and, besides, they help improve it for the next version and for other users. They will point out that no human beings have access to this information, that only machines and algorithms ever sift through it. They will have a point. But that does not make it any less creepy. Moreover, governments will inevitably want access to all that information as well. It will be too valuable to pass up.

Present-day users of smartphones and free web services have shown that they are willing to give up some degree of privacy in exchange for convenience. This is on the understanding that their data will be used to make profits – through targeted advertising, for example – and not abused to track their moves. So long as this compact holds, increasing surveillance is unlikely to bother everyday consumers.

But the future will require a more robust framework than exists today. Those that work in the field hold out hope that regulatory, legal and enforcement powers will keep large corporations, as well as overreaching government agencies, in check. The debate that started with Edward Snowden's massive leaks in 2013 is gathering momentum, and both companies and bureaucrats have pushed back against excessive data collection by states. Meanwhile, antitrust agencies around the world are keeping a check on big tech companies.

The second concern is that our world will be irredeemably mediated by corporations. To judge by the current state of consumer technology, it will be a small handful of firms that come to dominate the business of VR and AR. Every developer will be beholden to them; every consumer will have to agree to their terms and conditions. Their views of what constitutes acceptable behaviour and content (informed by the cultures they come from and by lawyers who want to limit liability) will form the basis of our interactions with the world. Indeed, just as they can disappear things from search results or social-network feeds, the power of AR may allow them to disappear people and objects from the real world too – it is all there but you just can't see it. Users who don't comply with the rules may find themselves cut off, cast adrift into a world with no reality but the real one. And the manner in which VR is evolving is very different from the development of the internet. Whereas the web was built on open standards and on the principle that anyone should be able to access, publish and link to each other, VR is being dominated by large companies with a fondness for "walled gardens".

Society will have to come up with ways to check the power of these companies, perhaps by enshrining new rights. Today, if Facebook or Google shuts down an account, its owner has little recourse. But as our online selves take shape, the question of who has the right to our avatars will become more urgent. Is signing up to a company's terms and conditions enough to sign away your virtual life? Or will companies have to allow their users to freely import and export their data? The latter seems more likely. Other rights will be up for debate too. What is your right to see the world unmediated, with naked eyes? Does that conflict with others' rights to block you or stay hidden? For a world based on virtual and augmented reality to work, these questions will need to be answered sooner rather than later.

A third worry is security. Despite years of work, computer security is nowhere near perfect. Even the most secure systems can be broken into by a determined hacker. And at the consumer level, security is only as good as any individual's ability to adhere to best practices. It is possible that by 2050 computer security will have advanced to a point where strong encryption, properly implemented, is the default, where passwords are a thing of the past and where it requires the might of a nation-state to break into systems. With the notable exception of Yahoo, the biggest tech companies have shown themselves up to the task of maintaining user security: hacking incidents at Facebook, Google, Amazon and Apple are notable by their absence. This may be one positive result of a VR world run by giants.

The final concern is perhaps the least worrying. Pessimists predict that VR will make the world a lonely place, with people absorbed in their private virtual worlds at the expense of the real world around them. VR will rot our children's brains, they fret. But these worries have been repeated over and over, about everything from social media, video games, television and rock music in the 20th century to the printing press in the 16th and indeed the written word in the time of Socrates (at least according to Plato).

Greybeards complain that children these days spend all their time staring into smartphone screens, but the kids in question are using their devices to engage with the world around them – taking pictures to Snapchat to their friends, Whatsapping about things they see, observing life in all its colour – rather than wandering around in made-up worlds. Even as technology changes, human beings remain fundamentally the same, and this means we will always want to interact with the world of flesh and flora.

Note

In addition to those mentioned in the text, I would like to thank the following people for their generosity with their time and insight: Justin Hendrix, Janet Murray, Alastair Reynolds, Mark Skwarek and Saschka Unseld.

PART 3

Megatech and society

13

The ethics of artificial intelligence

Luciano Floridi

The threat of monstrous machines dominating humanity is imaginary. The risk of humanity misusing its machines is real

SUPPOSE YOU ENTER a dark room in an unknown building. You may panic about some potential monsters lurking in the dark. Or just turn on the light, to avoid painfully bumping into the furniture. The dark room is the future of artificial intelligence (AI). Unfortunately, there are people who believe that, as we step into the room, we may run into some evil, ultra-intelligent machines. Fear of some kind of ogre, such as a Golem or a Frankenstein's monster, is as old as human memory. The computerised version of such fear dates to the 1960s, when Irving John Good, a British mathematician who worked as a cryptologist at Bletchley Park with Alan Turing, made the following observation:

> Let an ultra-intelligent machine be defined as a machine that can far surpass all the intellectual activities of any man however clever. Since the design of machines is one of these intellectual activities, an ultra-intelligent machine could design even better machines; there would then unquestionably be an "intelligence explosion", and the intelligence of man would be left far behind. Thus the first ultra-intelligent machine is the last invention that man need ever make, provided that the machine is docile enough to tell us how to keep it under control. It is curious that this point is made so seldom outside of science fiction. It is sometimes worthwhile to take science fiction seriously.

Once ultra-intelligent machines become a reality, they may not be docile at all but enslave us as a subspecies, ignore our rights and pursue their own ends, regardless of the effects that this has on our lives. If this sounds too incredible to be taken seriously, fast-forward half a century and the amazing developments in our digital technologies have led many people to believe that Good's "intelligence explosion", sometimes known as Singularity, may be a serious risk, and that the end of our species may be near if we are not careful.

Stephen Hawking, for example, has stated: "I think the development of full artificial intelligence could spell the end of the human race." Yet this is as correct as the following conditional: if the Four Horsemen of the Apocalypse were to appear, then we would be in even deeper trouble. The problem is with the premise. Bill Gates, the founder of Microsoft, is equally concerned:

> I am in the camp that is concerned about super intelligence. First the machines will do a lot of jobs for us and not be super intelligent. That should be positive if we manage it well. A few decades after that though the intelligence is strong enough to be a concern. I agree with Elon Musk and some others on this and don't understand why some people are not concerned.

And this is what Elon Musk, CEO of Tesla, a US carmaker, said:

> I think we should be very careful about artificial intelligence. If I were to guess like what our biggest existential threat is, it's probably that. So we need to be very careful with the artificial intelligence. Increasingly scientists think there should be some regulatory oversight maybe at the national and international level, just to make sure that we don't do something very foolish. With artificial intelligence we are summoning the demon. In all those stories where there's the guy with the pentagram and the holy water, it's like yeah he's sure he can control the demon. Didn't work out.

Just in case you thought predictions by experts were a reliable guide, think again. There are many staggeringly wrong technological forecasts by great experts. For example, in 2004 Gates predicted: "Two years from now, spam will be solved." And Musk speculates that "the chance that we are not living in a computer simulation is one in billions". That is, you are not real; you are reading this within the Matrix. Literally.

The reality is more trivial. Current and foreseeable smart technologies have the intelligence of an abacus: that is, zero. The trouble is always human stupidity or evil nature. On March 24th 2016 Microsoft introduced Tay, an AI-based chat robot, to Twitter. The company had to remove it only 16 hours later. It was supposed to become increasingly smarter as it interacted with humans. Instead, it quickly became an evil, Hitler-loving, Holocaust-denying, incestual-sex-promoting, "Bush did 9/11"-proclaiming chatterbox. Why? Because it worked no better than kitchen paper, absorbing and being shaped by the tricky and nasty messages sent to it. Microsoft had to apologise.

This is the state of AI today, and for any realistically foreseeable future. Computers still fail to find printers that are right there, next to them. Yet the fact that full AI is science fiction is not a reason to be complacent. On the contrary, after so much distracting and irresponsible speculation about the fanciful risks of ultra-intelligent machines, it is time to turn on the light, stop worrying about sci-fi scenarios, and start focusing on AI's actual and serious challenges, in order to avoid making painful and costly mistakes in the design and use of our smart technologies.

Pushing the envelope

One fundamental point needs to be understood to clarify such challenges. The success of AI is largely due to the fact that we are building an AI-friendly environment, in which smart technologies find themselves at home and we are more like scuba divers. It is the world that is adapting to AI not vice versa. Let's see what this means.

In industrial robotics, the three-dimensional space that defines the boundaries within which a robot can work successfully is defined as the robot's envelope. We do not build droids like *Star Wars*' C3PO to wash dishes in the sink exactly in the same way as we would. We envelop environments around simple robots to fit and exploit their limited capacities and still deliver the desired output. A dishwasher accomplishes its task because its environment is structured ("enveloped") around its simple capacities. The same applies to

Amazon's robotic shelves, for example. It is the environment that is designed to be robot-friendly. Driverless cars will become a commodity the day we can envelop the environment around them.

Enveloping used to be either a stand-alone phenomenon (you buy the robot with the required envelope, like a dishwasher or a washing machine) or implemented within the walls of industrial buildings, carefully tailored around their artificial inhabitants. Nowadays, enveloping the environment into an AI-friendly infosphere has started to pervade all aspects of reality and is visible daily everywhere, in the house, in the office and in the street. Indeed, we have been enveloping the world around digital technologies for decades without fully realising it.

In the 1940s and 1950s, the computer was a room and Alice used to walk inside it to work with it. Programming meant using a screwdriver. Human–computer interaction was like a somatic or physical relationship. In the 1970s, Alice's daughter walked out of the computer, to step in front of it. Human–computer interaction became a semantic relationship, later facilitated by DOS (disk operating system) and lines of text, GUI (graphical user interface) and icons. Today, Alice's granddaughter has walked inside the computer again, in the form of a whole infosphere that surrounds her, often imperceptibly. Human–computer interaction has become somatic again, with touchscreens, voice commands, listening devices, gesture-sensitive applications, proxy data for location, and so forth.

In such an AI-friendly infosphere, we are regularly asked to prove that we are humans by clicking on so-called CAPTCHA (the Completely Automated Public Turing test to tell Computers and Humans Apart). The test is represented by slightly altered strings of letters, possibly mixed with other bits of graphics, that we have to decipher to prove that we are a human not an artificial agent, for instance when registering for a new account on Wikipedia. Sometimes it is simply a box stating: "I'm not a robot." Software programs cannot click on it because they do not understand the message; humans find the task trivial.

Every day there are more humans online, more documents, more tools, more devices that communicate with each other, more sensors, more RFID tags, more satellites, more actuators, more data: in a word, more enveloping. And more jobs and activities are becoming digital in nature: playing, educating, entertaining, dating, meeting, fighting, caring, gossiping, advertising. We do all this and more in an enveloped infosphere where we are more analogue guests than digital hosts. This is good news for the future of AI and smart technologies in general. They will be exponentially more useful and successful with every step we take in the expansion of the infosphere. After all, they are the real digital natives. However, enveloping the world is a process that raises significant problems. Some, like the digital divide, are well known and obvious; others are more subtle.

A marriage made in the infosphere

Imagine two people A and H. They are married and they really wish to make their relationship work. A, who does increasingly more in the house, is inflexible, stubborn, intolerant of mistakes and unlikely to change. H is just the opposite, but is also becoming progressively lazier and dependent on A. The result is an unbalanced situation, in which A ends up shaping the relationship and distorting H's behaviour, in practice if not on purpose. If the marriage works, it is because it is carefully tailored around A.

In this analogy, AI and smart technologies play the role of A, whereas their human users are clearly H. The risk we are running is that, by enveloping the world, our technologies might shape our physical and conceptual environments and constrain us to adjust to them because that is the best or easiest – or indeed sometimes the only – way to make things work. After all, AI is the stupid but laborious spouse and humanity the intelligent but lazy one, so who is going to adapt to whom, given that divorce is not an option? You will probably recall many episodes in real life when something could not be done at all, or had to be done in a cumbersome or silly way, because that was the only way to make the computerised system do what it had to do. "Computer Says No", as the character Carol Beer in the UK comedy sketch show *Little Britain* would reply to any customer's request.

What really matters is that the increasing presence of ever-smarter technologies in our lives is having huge effects on how we think of ourselves and the world, as well as on our interactions among ourselves and with the world. The point is not that our machines are conscious, or intelligent, or able to understand or know something as we do. They are not.

There are plenty of well-known results that indicate the limits of computation, so-called undecidable problems for which it can be proved that it is impossible to construct an algorithm that always leads to a correct yes/no answer. We know that our computational machines satisfy the "Curry–Howard correspondence", for example, which indicates that proof systems in logic on the one hand, and the models of computation on the other, are structurally the same kind of objects, and so any logical limit applies to computers as well. Plenty of machines can do amazing things, including beating us at board games like chequers, chess and Go and the quiz show *Jeopardy!* The sky is the limit. And yet, they are all versions of a Turing machine, an abstract model that sets the limits of what can be done by a computer through its mathematical logic. Quantum computers, too, are constrained by the same limits of what can be subject to computation (so-called computable functions). No conscious, intelligent, intentional entity is going to emerge magically from a Turing machine.

The point is that our smart technologies, also thanks to the enormous amount of available data, some highly sophisticated programming and the fact that they can smoothly interact with each other (think of your digital diary synchronised across various platforms and supports), are increasingly able to deal with more and more tasks better than we do, including predicting our behaviour. So we are not the only agents able to perform tasks successfully, far from it. This is what I have defined as "the fourth revolution" in our self-understanding. We are not at the centre of the universe (Copernicus), of the biological kingdom (Darwin), or of the realm of rationality (Freud). After Turing, we are no longer at the centre of the infosphere, the world of information processing and smart agency, either. We share the infosphere with digital technologies.

These are not the children of some sci-fi ultra-intelligence, but ordinary artefacts that outperform us in ever more tasks, despite being no cleverer than a toaster. Their abilities are humbling and make us re-evaluate our human exceptionality and our special role in the universe, which remains unique. We thought we were smart because we could play chess. Now a phone plays better than a chess master. We thought we were free because we could buy whatever we wished. Now our spending patterns are predicted, sometimes even anticipated, by devices as thick as a plank.

What does all this mean for our self-understanding? The success of our technologies largely depends on the fact that, while we were speculating about the possibility of ultraintelligence, we increasingly enveloped the world in so many devices, sensors, applications and data that it became an IT-friendly environment, where technologies could replace us without having any understanding, intentions, interpretations, emotional states, semantic skills, consciousness, self-awareness or flexible intelligence (as in seeing a shoe as a hammer for a nail). Memory (as in algorithms and immense datasets) outperforms intelligence when landing an aircraft, finding the fastest route from home to the office, or discovering the best price for your next fridge.

So smart technologies are better at accomplishing tasks, but this should not be confused with being better at thinking. Digital technologies do not think, let alone think better than us, but they can do more and more things better than us, by processing increasing amounts of data and improving their performance by analysing their own output as input for the next operations, so-called machine learning. AlphaGo, a computer program developed by Google DeepMind, won the board game Go against Lee Sedol, the world's best player, because it could use a database of around 30 million moves and play thousands of games against itself, "learning" each time a bit more about how to improve its performance. It is like a two-knife system that can sharpen itself. Yet think for a moment what would have happened if the fire alarm had gone off during the match: Sedol would have immediately stopped and walked away, while AlphaGo would have been calculating the next move in the game.

So what's the difference? The same as between you and the dishwasher when washing the dishes. What's the consequence? That any apocalyptic vision of AI can be disregarded. The serious risk is not the appearance of some ultra-intelligence, but that we may misuse our digital technologies, to the detriment of a large percentage of humanity and the whole planet.

Beware of humans

We are and shall remain for any foreseeable future the problem, not our technology. This is why we should turn on the light in the dark room and watch carefully where we are going. There are no monsters, but plenty of obstacles to avoid, remove or negotiate. We should be worried about real human stupidity, not imaginary artificial intelligence. The problem is not HAL but H.A.L. – humanity at large.

Thus we should rather concentrate on the real challenges. By way of conclusion, I will list five of them, all equally important.

- We should make AI environment-friendly. We need the smartest technologies we can build to tackle the very concrete evils oppressing humanity and our planet, from environmental disasters to financial crises, from crime, terrorism and war to famine, poverty, ignorance, inequality and appalling living standards. For example, more than 780 million people do not have access to clean water and almost 2.5 billion do not have access to adequate sanitation. Some 6 million–8 million people die annually from the consequences of disasters and water-related diseases. This, not AI, is among "our biggest existential threats".

- We should make AI human-friendly. AI should be used to treat people always as ends, never as mere means, to paraphrase Immanuel Kant.

- We should make AI's stupidity work for human intelligence. Millions of jobs will be disrupted, eliminated and created. The benefits of this transformation should be shared by all, and the costs borne by society, because never before have so many people undergone such a radical and fast transformation. The agricultural revolution took millennia to exert its full impact on society, the Industrial Revolution took centuries, but the digital one took only a few decades. No wonder we feel confused and wrong-footed.

- We should make AI's predictive power work for freedom and autonomy. Marketing products, influencing behaviours, nudging people, or fighting crime and terrorism should never undermine human dignity.

■ Finally, we should make AI make us more human. The serious risk is that we may misuse our smart technologies. Winston Churchill once said that "we shape our buildings and afterwards our buildings shape us". This applies to the infosphere and the smart technologies inhabiting it as well. We'd better get them right, now.

14

The data-driven world

Kenneth Cukier

The ubiquitous application of data on a massive scale and in innovative ways will make a lot of things easier, cheaper and more abundant

YOU WAKE UP and your personal robot hovers through the air to serve breakfast in bed: a protein pill and espresso lozenge. Just the act of yawning lets the breath sensor in the ceiling analyse the data to test your biochemistry for any ailment. Your self-piloting jetpack flies you to work.

Actually, dream on. This is science fiction. But over the next three decades, as the techniques of artificial intelligence (AI) worm their way into all areas of life, fundamental changes are under way. Every aspect of business and society will be touched by data, just as they have been by computing and the internet over the past three decades.

The modern revolution in the sciences dates from Galileo's *Discorsi* in 1638, where he put forward the idea that all natural phenomena could be expressed in the language of mathematics. So, too, a new era is emerging when all activities under the sun will be better understood and optimised by collecting and analysing the data these activities generate. Information will be regarded as an essential resource. The 19th century was powered by steam, the 20th by oil, and the 21st century will be fuelled by data.

With advanced AI techniques, we can learn both at an unprecedented scale and in an automated way. Kevin Kelly, an American philosopher of technology, calls it to "cognitise" – that is, to inject intelligence into everything we do. The ability to do so is thanks to tiny computer chips and sleek algorithms, yet the actual intelligence depends on data. And because data are the lifeblood, the gadgets will not just use data but constantly collect new data too.

We see the precursors of this already, in things as rudimentary as thermostats (a Google product called Nest) and the bevy of fitness trackers that count our steps and heart-rate, to say nothing of magical, always-on, voice-activated personal assistants (Google Home and Amazon Echo). By 2050 these will be as common as wristwatches or radios a half-century ago.

Applying data to our everyday activities is another way of saying that we bring to bear empirical evidence of how the world works. Society has long done this, of course – but when the size of data was limited, all that was knowable were big, unmissable patterns. Now, as so much more data become available, we can detect patterns that are far more subtle. If the same sorts of efficiency improvements take place over the next 35 years that we have experienced over the past 35 years from computers, we can glimpse the basic outline of what life in 2050 will be like.

It breaks down into three main trends. First, things that are hard to do today will become easier. Second, things that are costly to do will be cheaper. Third, things that were scarce will be more abundant. In short: easier, cheaper, abundant. Let's take these three trends and map them onto some of the biggest areas of society: health care, education and law.

Doctor, heal thyself

Today the practice of medicine has more in common with the 19th century than with the 21st. Doctors rely on their memory of medical-school textbooks and years of accumulated experience to make decisions. This sounds perfectly sensible. But in fact it is ludicrous: it is impossible for any one practitioner to be familiar with all possible conditions and the best course of care, especially as new treatments are constantly emerging.

If Google can rank the most relevant webpage among billions of candidates and Amazon can uncannily recommend your next purchase, shouldn't doctors rely on a computer for every diagnosis? By 2050 this will probably become standard. Medical records will be electronic and algorithms will comb through all the data to spot correlations of best practice and adverse side effects.

The database will be the smartest doctor in the world: it remembers every case and sees patterns among treatments and outcomes to know what works best in a given situation. Doctors may still make the final decisions. But they might open themselves up to malpractice suits if they try to make a diagnoses *without* tapping into big-data systems to corroborate their judgment – just as airline pilots would lose their job today if they were to turn off the autopilot, slip on a leather helmet and goggles, and attempt to land the plane "old school".

IBM's cognitive-computing platform, Watson, is already ingesting medical information in the hope of being able to diagnose illness, as a support for doctors. Big-data systems are even now being used for designing new drugs. And robotic surgical systems are "trained" on the accumulation of data from past operations, much as a self-driving car relies on the experience of previous driving situations.

One domain where big data will surely start to upend health care is computational pathology. In 2011 a team of researchers led by Andrew Beck at Harvard University used computer-vision and a machine-learning algorithm to analyse biopsies of breast-cancer cells along with the patients' survival rate, to see if the system could do as well as humans in predicting cancer. Astoundingly, it did. More than that: among the 11 traits that the algorithm used to predict that a biopsy was highly cancerous were just eight that referred to the cells themselves. Three of the traits related to the surrounding "stromal" tissue, things that doctors didn't know to look for. It was hidden to the human eye, but uncovered from analysing mountains of data.

The technology is still in the lab, and regulations need to be revised to allow computational pathology. Yet by 2050 it will be the way medical diagnostics happens. The point is that data will revolutionise the delivery of health care. More broadly, anything that requires highly specialised training, judgment and decision-making under conditions of uncertainty will be done better by an algorithm than by a human. It will be more accurate, faster and cheaper.

Teaching the teachers

A second area ripe for being transformed by data is education. When public education became common in the West in the 1800s, it was meant to replace a world of private tutors, which was how the upper classes had been educated. Teaching had been individually tailored around a person's abilities. But the education system mirrored the industrial organisation at the time, the factory. Instruction was mass-produced. Students rolled down an assembly line; teaching was one-size-fits-all. It was hard to do it any other way.

Data were used then, as now, as occasional snapshots in time: a test score here, a paper grade there. But they were not collected and analysed in a continuous fashion to learn what worked best or to tailor the instruction to the needs of the student. Until recently, it would have been too costly and cumbersome to do so. Yet these constraints are going away. As a result, we can reimagine what education will look like in 2050: data will be constantly used to track student performance – and teacher performance – to learn what works best for learning. And the data will enable the reintroduction of individually tailored, customised instruction that was lost in the era of mass-produced public education.

The platform of education will be digital, so data can be collected at every turn. In some instances, this will mean a "flipped" classroom, whereby students hear the lectures at home (instead of doing homework alone) and come to class for the problem-solving exercises (when the teacher is there to help).

There are already early examples of this. When Andrew Ng, a computer-science professor at Stanford, taught an online course in 2012, data helped him improve the teaching. While analysing the video viewership logs, he noticed a strange pattern: students tended to watch sequentially, but around Lesson 7 they jumped back a few weeks to Lesson 3. With tens of thousands of students in the online class, the pattern was not a fluke. What was happening? He looked into it and realised that the earlier class was a maths refresher course. Several weeks later, the students were getting hit with complicated material and didn't feel confident in their skills. Ng therefore knew he needed to change the curriculum to include more maths classes at the start so that students would be prepared for the harder stuff.

Online classes are just the start. When a textbook is an e-book, the device learns from the student as much as the student learns from the device. The device knows if the student is reading, and how quickly. If the student's attention wanders (detected by a slower reading rate), the e-book can automatically insert a short video explainer or a pop quiz to revive the student's attention. The e-book will know if the student is reading on a Sunday afternoon at home or a Monday morning on the school bus. It will be able to see whether higher test scores are correlated with studying before or after dinner.

Thus data in education, as in health care, go from a stock to a flow: from something gathered in one-off, discrete bits, to something constantly collected, a steady stream. This will allow a technique called adaptive learning to take hold. The idea is to analyse student actions and choose the material and pace of instruction that are exactly right for the individual. So if a student aces three maths problems about triangles in rapid succession, the software knows to move on to something more challenging. If they struggle with circumferences, it adds extra questions. Today, even A-grade students have "Swiss cheese" gaps in their knowledge; adaptive learning can ensure that they master all areas of a subject before moving on.

By 2050 the data will enable education to return to its roots by tailoring the teaching to fit the individual, not the other way around as happens today. Education becomes easier to do, cheaper to offer and more abundant, in that the system can serve more students.

Kill the lawyers

The rule of law is precious but most people would probably wish for a world with fewer legal professionals. "The first thing we do, let's kill all the lawyers," proclaims Dick the Butcher in Shakespeare's *King Henry VI, Part Two*. Until recently the law has been one of the sectors least touched by data; it is a domain of written arguments and human judgment. But this is changing. By 2050 data will be at the core of the legal profession and indeed the concept of justice.

Already, data are used to identify unbalanced police activity and sentencing, such as the frequency with which young black men in the US are frisked or convicted compared with white men. Several firms offer jury-selection services to trial attorneys, providing data on the likelihood that, say, an Asian woman will vote to convict a female defendant, when the legal team is vetting potential jurors. The system works by correlating mountains of demographic data on jurors with trial outcomes, allowing lawyers to improve the odds that they make a good pick.

Another area is called e-discovery. Complex corporate litigation may involve the need to review millions of pages of documents. In the past, armies of young lawyers culled the material. It was expensive and inefficient. But algorithms can scan for keywords or map e-mail traffic to reveal suspicious behaviour – in a fraction of the time and at a fraction of the cost, and with better accuracy than humans.

Other data-intensive practices are on their way. A start-up called Ravel has placed mountains of US case law – motions, orders, verdicts and the like – into a massive data warehouse in order to extract details and find correlations. The aim is to revolutionise legal research. Eventually lawyers will not only find relevant precedents (as current systems do), but also see which precedents are more commonly cited in winning versus losing cases, down to the level of the district or judge. They will be able to identify the most common arguments the opposing counsel uses, and what are the most effective strategies against them.

By 2050, most legal briefs and contracts will at least initially be drafted by an algorithm tapping into mountains of data to find what is most advantageous. But the documents will need to be reviewed and approved by a human attorney to make sure that the claims are ones the client and counsel stand by. Likewise, although an algorithm could produce reliable and consistent verdicts for basic disputes, the bar associations will still require a human judge so the parties and the public have at least the feeling that the specific merits of the case are being weighed. Just as a staple of jurisprudence is that justice not simply be done, but that it be seen to be done, so too the new thinking will be that everyone deserves a day in court – with a human, not a bot.

Data will improve the administration of law, and create a more just society. By 2050 there will never be a case of bail being denied to a defendant because a judge has an inkling that the fellow is a flight risk. Instead, the amount will be based on data, just as banks now determine mortgage rates using data. Similarly, the idea that members of a parole board can look a prisoner in the eye and know if he can be released early without being a threat to society will be understood as the fiction that it is. Instead, the decision will be based on the statistical likelihood that the prisoner will reoffend.

One effect of the fusion between data and law is that the legal profession will become more efficient, with better arguments and faster resolutions. (Whether clients will get lower bills remains to be seen.) With big data, law will be easier and cheaper. It will also be more abundant, in that access to justice is expanded. Today, the idea of turning to the judicial system to settle a dispute is a luxury not a right. Many people are shut out – the law simply cannot help them because the mistreatment does not reach the level where the resolution is worth the rigmarole. The cost of seeking redress through the law may decline and the ability to provide representation will expand as data change the economics of legal services. We may find that the rule of law becomes broader and deeper in society, much as the cost of storing and transmitting information fell because of the printing press, enabling the dissemination of knowledge to expand.

A job apocalypse?

Throughout society, as with medicine, education and law, data will produce vast improvements. Yet there are legitimate concerns that this might take a huge toll on jobs. If an algorithm can detect cancer better than human pathologists, won't many medical practitioners be out of work? If a single professor can teach hundreds of thousands of students simultaneously online, might we need fewer instructors? And perhaps there is no need to kill the lawyers – they will cull themselves, as only a handful of senior partners are required to look over the work of the algorithms, not rows of junior associates.

It will surely be a rough transition for many. In the longer term, though, there is no reason why the market for services like medicine, education and law should not expand and employment even increase. The nature of jobs may change, too, and for the better, as algorithms handle the skull-drudgery that no one wished to do anyway.

Take pathologists, for instance. Some will still be needed to interact with the algorithm, check its work and make sure that new findings in the field are accurately updated in the technical system. As efficiency improves and costs plummet, instead of examining cell samples only when someone feels a lump and goes along to the doctor's surgery, analyses will be performed every day from data collected by regular activities. And we will do this not just for those who can afford it but for the entire population – no doubt learning new things about the progression of disease that we never knew before when we lacked the data. In such a world, we may need more pathologists than before, not fewer.

The pathologists' skills will need to change, to capitalise on their humanity. More than today, they will need to be able to interpret and explain the output of the algorithmic diagnosis to patients going through a health crisis. Pathologists will have to improve their bedside manner. Medical schools will need to teach communications and psychology alongside the biology.

Other professions may experience a similar boom in employment. If classrooms are flipped, hands-on teaching will become even more important. We will just get rid of the sage on the stage who drones on while students sleep. Teachers will be like sports coaches – people who find the right balance between pushing players to excel and picking them up when they fall short. But this will require new skills. Likewise, we may need more lawyers – and then we will have a more just society, as the algorithms work with attorneys to expand access to law to more people more of the time.

The arrow of causality, bent

These advances from data will come with a twist: we will have a better understanding of "what" is happening in the world but a worse understanding of "why". The machine-learning system identifies the cell sample is not cancerous but cannot indicate why – the patterns are so numerous and subtle they are outside human comprehension. The educational algorithm identifies someone as at risk of dropping out, but none of the features alone explains why it should be correct. The software tells police that they should patrol a specific block because it is likely to experience an increase in crime, but it cannot indicate why.

So we replace the vagaries of life before big data, in which our knowledge was stunted for lack of information, with newer vagaries of life in an age of AI, in which we have a surfeit of information but understand the reasons less. Society gets a victory in more efficiency, but the trade-off is that it lacks knowledge about the cause and effect behind the system. A big part of how decisions will be made in 2050 will be a black box, devoid of the transparency that is a cornerstone of accountability.

Regulation will have to adjust to this new world. The EU's data-protection directive that comes into force in 2018 points towards giving the public a "right to explanation" for algorithmic decisions, just as it has a "right to be forgotten" for privacy. US regulators, for their part, have openly worried that advanced data processing may lead to new forms of high-tech discrimination. The question of causality is the rub. A lot is at stake: an engineer at a large US medical-device company admits that one of its implanted devices uses a slightly inferior method because the better algorithm is based on "deep learning" and fails to provide a clear explanation of how it works, as regulators require.

By 2050 the world will have become comfortable with trading causality for efficiency, in the same way that by the end of the Enlightenment society accepted that what could be visually observed (like the sun revolving around the Earth) did not explain natural phenomena. Data will be humbling for humanity.

Data, data, everywhere

As the value of data increases, there will be a call to put a clearer monetary value on, in particular, personal information. It will be tempting to make personal privacy a property right, in order to better protect it. Firms will need to get clearer permission to use personal data and suffer greater economic damages if they fail to protect privacy or misuse the data.

Just as we have banks for monetary assets, a new sector of the economy will emerge, the data bank, for the informational assets of companies and people. In 2016 many web surfers downloaded ad-blocking software that sits between the device and the website to block online ads. It is easy to imagine that, in future, similar software will sit between the person and the website to manage the data transfer, for a fee. However, hitherto free services like Facebook and Google will come with a price tag unless we are willing to cough up the data. So privacy in 2050 will probably be a luxury good, like flying business class or owning a second home.

By then, data will be brought to bear in almost everything that happens. This will produce three major changes. First, what we do will be more efficient, and in some cases done in an entirely new way. Second, we will have a better understanding of the world as it truly is, not just what we could observe with a smidgeon of data. Third, the role of data will go from a stock to a flow: that is, we will track things constantly like a motion picture, not just occasionally like a frozen snapshot in time.

Big data will not bring heaven on Earth. AI will not eliminate mortality. The lion will not lie down with the lamb; nor will we pound our Kalashnikovs into selfie sticks. But almost everything will be optimised by data. The world will be better for it.

15

Imagine all the empowered people

Melinda Gates

If every woman in the world had a smartphone, it would transform their lives

"WHO IS SABITA?" I was looking right at Sabita Devi when she said these words. She was describing her life as a wife and mother in Jharkhand, one of the poorest states in India, where she has spent most of her days inside the four walls of her home. "No one in my village knew my name," Sabita told me. Her contact with the outside world was mediated entirely by her husband: who she could talk to, what she could buy, when (and if) she could see a doctor. She was isolated from everyone and everything but her children.

That feeling of being alone and apart is common to women across space and time – just ask any American feminist author over the past century. The solution is frighteningly complex. It involves gradually changing prevailing norms in thousands of cultures, and there is no app for that.

But that doesn't mean that apps won't help – a lot. Suppose every woman in the world had a smartphone. That could help shatter their isolation and unleash their powers like never before.

Take health care. When every woman has a smartphone, she will be able to get the right information, at the right time, in the right format. For example, an illiterate Nigerian woman in her first trimester of pregnancy might receive a voice message in the Hausa language describing anaemia and explaining how to get iron supplements. The same system can remind her when she needs to go for a prenatal visit, or take her children for immunisations. Eventually, she'll be able to use her phone for a video consultation with a doctor when non-routine problems arise. With the doctor's guidance, she'll also be able to use her phone to check her temperature, blood pressure and other vital signs to assist with her own diagnosis.

Then there's farming. Poor farmers stay poor in large part because they don't have the information they need to get the most out of their land. For instance, they know almost nothing about the nutrient content of their soil, which means they can't choose the right fertilisers or the crops that will grow best. They also don't have reliable information about market prices, so they are forced to take the price being offered by whichever trader comes by. Since most farmers in Africa (and many in South Asia) are women, this is a women's problem. And women farmers are also less productive than men, on average, for reasons ranging from gender bias in agriculture training to how hard it is for women to hire and manage male labourers during the harvest.

But with smartphones, women farmers will be able to watch videos of local farmers providing training based on local soil and weather conditions. They will also have apps that tell them what prices their crops are fetching at various markets, so they can be informed sellers. Using their phones to connect to each other, women farmers can also organise effectively in co-operatives so they can express their demands as members of a powerful group rather than as isolated individuals.

Banking provides another example. Even the poorest women have assets. One of the keys to empowerment is making sure they can control what happens to these. The most striking study I've ever read shows that when women determine the family budget, children are 20% more likely to survive – simply because of the things they tend to spend money on (such as food and health care).

Traditional banks haven't found it profitable to serve customers making transactions in tiny amounts. That leaves poor women to navigate the informal economy, hiding cash, buying jewellery or livestock as an illiquid form of savings and borrowing from moneylenders at usurious rates.

Digital technology slashes transaction costs, though, which means that people can save and borrow money or purchase insurance securely and in small amounts through their phones. This technology is already taking hold in countries like Bangladesh and Kenya, but many emerging digital economies are male-dominated because men own mobile phones at much higher rates than women. In Bangladesh only 46% of women own a phone, compared with 76% of men. Consequently, only 13% of Bangladeshi women have used mobile money, compared with 32% of men. When these percentages even out around the world, it will unlock the economic prowess of a billion people.

It isn't hard to do

The beauty of this vision is that you don't need to strain that hard to see it clearly. I am not positing a science-fiction future. More than two-thirds of the people on the planet have access to mobile phones, and, increasingly, these are smartphones. Last year, more than 1 billion smartphones were sold worldwide.

Those statistics are encouraging, but we're not even close to universal smartphone coverage. Data must get cheaper. Many people with smartphones don't even use the internet because it is too expensive. And connectivity has to reach the most remote places in the world, where the need for connection – in the widest sense of the word – is greatest.

That still leaves the job of making sure that women own phones in the same numbers as men – and that there are apps designed to address their specific needs. An even larger hurdle is the fact that illiteracy is becoming a girls' problem, and women who can't read will never get the most out of their mobile phones.

No technology by itself can usher in gender equality. But, by helping women forge connections they've never had, the smartphone can make a big difference.

Note

This article first appeared online in *The Economist*'s collection of scenarios, *The World If*, in 2015. It is published here in print for the first time.

16

Megatech versus mega-inequality

Adrian Wooldridge

Technology has been responsible for much of the rise in inequality – but it could also be a big part of the solution

IN 1845, at the height of the first Industrial Revolution, a young politician on the make published a popular novel with the double-headed title of *Sybil, or The Two Nations*. Benjamin Disraeli provided his readers with the requisite amount of romance in the story of Sybil. His real subject, however, was the division of the UK, the engine room of the new economy, into two nations:

> *Two nations; between whom there is no intercourse and no sympathy; who are as ignorant of each other's habits, thoughts and feelings as if they were ... inhabitants of different planets; who are formed by a different breeding, are fed by a different food, are ordered by different manners, and are not governed by the same laws: the rich and the poor.*

Disraeli's two-nation problem is with us again, and not just in the UK. Almost all the productivity gains of the past 30 years have been gobbled up by the richest 1%. Growing inequality is dividing society into two worlds, "between whom there is no intercourse and no sympathy". The cosmopolitan elites deride the parochial masses as "racists", xenophobes and troglodytes. The parochial masses denounce the cosmopolitan elites as self-dealing traitors. Growing inequality is undermining popular faith in the basic pillars of the capitalist order: opportunity and upward mobility, economic growth and shared prosperity. The UK's decision to leave the European Union in June 2016 and the election of

Donald Trump as US president in November 2016 were just the most dramatic examples of a populist hurricane that is being whipped up by worries about inequality and threatens to wreck the liberal order.

There is a growing consensus that "something must be done" to calm this hurricane. Barack Obama dubbed declining rates of social mobility as "the defining issue of our time". Hillary Clinton pronounced that "we don't have enough social mobility". Donald Trump claimed that he championed ordinary Americans against corrupt and complacent elites. Commentators of all political persuasions agree. On the right Charles Murray has written a book called *Coming Apart*. On the left George Packer has written an equally despairing tome called *The Unwinding*. In the centre David Brooks warns that inequality represents capitalism's "greatest moral crisis since the Great Depression".

But what exactly is that "something"? Liberals routinely mention education. Yet improving schools is remarkably difficult: since 1970 real (ie, inflation-adjusted) spending per pupil in US public primary and secondary schools has more than doubled while test scores have remained more or less flat. Clever initiatives such as charter schools and magnet schools frequently succeed for a while but then falter. Conservatives increasingly point to immigration control. But erecting barriers to the flow of people can reduce growth and dynamism; talent, like capital, will go to places where it is welcome.

The contention of this article is that a goodly part of the solution lies in technology.

Divide and school

Technology, it is true, is responsible for much of the increase in inequality. Even places, such as Japan and the Scandinavian countries, that have adopted more egalitarian social policies than the Anglo-Saxons are seeing rises in inequality. Technological innovation is benefiting two sorts of people. Clever knowledge workers can use computing power to improve their output while reducing their support staff: star academics can write more papers, star lawyers can digest more material, star journalists can sniff out better stories. The cleverest knowledge workers can produce breakthrough ideas that they can sell in a global market: build a better electronic mouse and the world will beat a path to your door. At the same time innovation is reducing the bargaining power of anybody who performs routine functions: any job that can be reduced to a repeatable formula will eventually be taken over by an intelligent machine. A broad swathe of middle-class workers are suffering from the same fate as industrial workers did from the 1980s onwards – a relentless downward pressure on their wages, which will give way inexorably to the destruction of their jobs. Zero-hours contracts are zero contracts in the making.

Technological disruption has turned San Francisco, the capital of the current industrial revolution, into one of the most polarised cities in the US. The city bends over backwards to cater for the peccadilloes of the tech elite. Cafés specialising in Asian-Mexican fusion foods, or whatever the latest food fad dictates, are crowded with the industry's finest, who can enjoy not just the food but free "device-charging" pads. But people who don't work in the tech economy are being priced out of their old homes as rents soar. And the homeless multiply daily, leaving a smell of sweat and urine wherever they go. Young techies, their ears permanently blocked by headphones, pick through the litter and detritus of the country's failures.

Yet the very force that is dividing society can also, if cleverly directed, boost opportunity. Information technology has a levelling dimension: it empowers consumers, challenges rent-seeking elites and pushes down the cost of many services. The average smartphone contains as much computing power as was available to the whole of MIT in the 1950s. It also has a universalising dynamic: it reaches into the farthest corners of the world, taking the luxuries of the rich and turning them into commodities for the poor. Joseph Schumpeter once remarked that the "capitalist achievement" does not lie in "providing more silk stockings for queens but in bringing them within reach of factory girls". The same is even more true of devices like personal computers and smartphones. Ned Ludd thought that the solution to the "machinery question" was to smash the machines. A more intelligent solution is to turn those machines into your servants.

One of the biggest tech trends of the next 50 years will be an optimistic one: policymakers will learn how to harness technology to address the problem of excessive inequality. Indeed, there are already plenty of examples of this happening.

The most obvious thing that technology can do to promote equality is to reduce the costs of providing labour-intensive services such as education. The cost of providing high-quality education has risen faster than the cost of living for decades. Elite institutions have become luxuries for the rich: the average income of parents of Harvard students now surpasses $450,000. Less elite institutions have seen their quality erode. Though continental European countries pride themselves on making universities free and open to all, not a single euro-zone university makes it into various lists of the world's 30 best universities. Many are little more than academic slums.

William Baumol once suggested that there is something inevitable about this: productivity growth in the service sector is bound to be slower than productivity growth in the manufacturing sector because it always takes four people to play a string quartet. But Baumol's own example undermines his case. Thanks to technology, you can listen to an almost perfect reproduction of a string quartet in your home without the annoyance of a trip to the concert hall and the coughing and spluttering of fellow audience members; indeed, thanks to Spotify and Apple Music you can listen to most of the music that has ever been commercially released for a relatively reasonable monthly fee. The digital revolution has transformed large sections of the service sector, such as retailing, and the intellectual sector, such as journalism. It will inevitably transform education and medicine by applying similar labour-saving techniques.

Yes we Khan

Technology is already attacking costs. The Khan Academy currently provides free lessons to more than 4 million children a month and is constantly adding to its library of more than 5,000 lessons. Bill Gates encourages his own children to use the academy. In the US a tenth of university students now study exclusively online and a quarter do so some of the time. Leading universities such as MIT, Stanford and the University of California at Berkeley have put some of their lessons online. The University of the People offers free higher education (not counting the few hundred dollars it costs to process applications and mark exams). It is certainly true that the dropout rate from MOOCs (massive open online courses) has been disappointingly high. But we are still at the early stages of developing new courses. Online universities will inevitably develop more sophisticated ways of providing social support and encouragement for their students, much as the UK's Open University did in the 1960s by mixing television-based instruction with residential periods. And the power of technology is growing dramatically: online videos are becoming more vivid by the month, and it is only a matter of time before we are able to project holograms of star professors into lecture theatres around the world.

Technology is also being used to address questions of efficiency. Numerous companies, such as Reasoning Mind and DreamBox, produce "adaptive" or "personalised" learning machines that gather data on the individual performance of children and then automatically tailor teaching to their needs. These programs are particularly popular in charter schools such as Rocketship Education in California and New Classrooms in New York. A growing number of schools are using iPads to "flip the classroom": instead of focusing on transmitting knowledge during lessons, teachers encourage children to absorb knowledge at home, via iPads, and spend lesson time discussing what they have learned.

One reason inequality is so hard to tackle is that it starts so early on, in the womb, the pram and pre-school. Middle-class mothers make more effort to provide their children with a healthy environment in the womb. Middle-class children routinely hear millions more words spoken in their first two years than working-class children do. Middle-class parents are more likely to provide their children with pre-school education. Policymakers initially tried to tackle these problems by fairly obvious interventions such as enriched pre-school education. Now they are trying more subtle interventions such as nudging mothers to eat more healthily or to stimulate their children mentally. New technology can do a lot to make such subtle interventions easier. In 2014 the mayor of Providence, Rhode Island, Angel Taveras, invented an early-childhood intervention programme, Providence Talks, in which parents wear devices that record how many words they say daily and receive regular advice on how better to talk to their children. Taveras's device is the beginning of a new trend: within a few years welfare departments will routinely provide poor mothers with devices that provide them not only with advice about bringing up their children but also with measures that help them see how well they are doing.

Hey, Jude, don't be unpaid

Another reason for inequality's persistence is the Jude the Obscure problem: poorer bright children languish, ignored and unrecognised, while indifferent middle-class children thrive, pushed to the front of the queue by doting parents. Modern technology promises to bring much more rigour to selection: schools and universities will be able to use ever more sophisticated techniques – and ever bigger volumes of data – to find talent that would once have been undiscovered. The Israel Defence Forces provides a glimpse of the future. Unit 8200, an elite cyber-security group, combs Israel's schools for the best talent regardless of social background. Scouts monitor children's performance in video games as well as more routine academic tests for signs of genius. And once they have spotted their geniuses they provide them with rigorous training and generous scholarships. Graduates of the unit have founded a host of technology companies, such as CheckPoint, Imperva, Nice, Gilat, Waze, Trusteer and Wix.

Technology can also address two of the biggest weaknesses of modern education: vocational education and occupational guidance. Vocational education is the poor relation of modern education: schools' traditional obsession with academic education has only been deepened by the expansion of the universities. Technology can bring glamour and energy to vocational education. For example, some companies now provide high-tech job training that allows pupils to see what it is like to control powerful machines or perform delicate operations.

Occupational guidance has historically been a slapdash affair: when I was at university careers advice meant going to see a retired army officer in the wastelands of north Oxford. Now innovators are beginning to apply technological solutions to the problem. LearnUp, a Silicon-Valley-based company, tries to provide jobseekers who lack college degrees with the skills that they need to get jobs. Alexis Ringwald, the company's co-founder, gave up a conventional Silicon Valley career to spend six months studying the other Bay Area, the one where people without college education spend months idling on unemployment lines, and sometimes end up sleeping in doorways. She found that many long-term unemployed people lacked basic skills: they didn't bother to find out what jobs entailed or how to present themselves in interviews. LearnUp has created an online job coach that teaches people basic skills such as how to behave during interviews and how to use a photocopying machine. It is now forming partnerships with big companies that suffer from persistent skills shortages. For example, it provides free online job-skills preparation for an estimated 200,000 jobseekers at 350 Old Navy stores across the US.

Several technology firms, among them Cisco, are drawing up plans to produce "vocational guidance machines" that will be able to provide young people with advice about what they might want to do, on the basis of their abilities and aptitudes, and where they might want to go to get education and training, on the basis of what is available both physically and virtually. But this is only the beginning of an explosion of activity. Why not invent a Tinder for academic mentors? That is, an app that would introduce you to thousands of people who are willing to act as mentors or home tutors, either on screen or in person. Why not apply the power of gaming technology to learning? Poor schoolchildren, particularly boys, might well overcome their aversion to learning if they saw it as an extension of the gaming world. It's worth a serious try. And why don't universities get as serious about looking for bright working-class students as they do about looking for athletic superstars? Universities routinely send scouts around the country to look for talented children. Why not use electronic spies to find children who have a genius for highly complicated video games, as Israel's army does?

Technology can also be used to improve the lives of rank-and-file workers rather than marginalising them, as many fear. General Electric claims that it is more interested in using technology to "upskill" workers than to replace them: by providing semi-skilled workers with iPads and the appropriate apps it can set them on tasks that used to be reserved for much more skilled workers. It also points out that its habit of offering prizes to people outside the company who can solve problems has increased the chances that people who live far from its normal recruiting ground will end up with a job.

You say you want a data revolution

Social inequality is reinforced by health inequality. Poorer people die earlier than richer ones and live more unhealthy lives. Big data can be used to address public-health problems more quickly than in the past. Such troubles are more prevalent among the poor than the rich. Wearable devices can diagnose problems before they occur, remind forgetful patients to take their drugs and monitor patients after they have left hospital. The cheaper and more pervasive these devices become, the more they will reach the poor as well as the rich.

Social inequality is also reinforced by crime and corruption. Body cameras can keep rogue policemen in line or get them prosecuted. Security devices can reduce car crime. Large police departments such as the NYPD and the LAPD routinely use big data to allocate police to areas where crimes are occurring. Corruption is one of the main problems in the emerging world (and in some rich countries as well). Technology can help bring it to light. The victims of bribery can record corrupt officials. Whistle-blowers can publish examples of shady dealing, as they did with the Panama Papers. The biblical adage that "you may be sure your sins will find you out" is doubly true in a world of ubiquitous cameras and big data.

It is important not to fetishise technological innovation. Schumpeter may have been right that innovations inevitably trickle down to the masses. But how quickly this happens and how comprehensively depends on public policy. Some countries, such as Sweden and Singapore, have done much better than others at making fast broadband ubiquitous. Some technology-enabled initiatives, such as Grameen's scheme to equip micro-entrepreneurs in Bangladesh, pay off handsomely. Others, such as Nicholas Negroponte's one-laptop-per-child initiative, fizzle out. Technological progress will not solve the problem of rising inequality on its own: the rich are usually faster at adopting new technology and better at using it to upgrade their skills rather than, say, entertain themselves. Dysfunctional communities will not become functional just because they are connected to the internet. When I mentioned the idea of giving children iPads, so that they could continue to learn at home, a teacher in a poor school in Johannesburg rolled her eyes and pointed out that they would be stolen and sold in an instant.

But technology provides an immensely powerful tool if it is harnessed by enlightened policymakers and connected with a broader social policy of addressing inequality. Policymakers need to start by thinking carefully about two questions. What does equality mean? And who is responsible for delivering it?

Smart technology, wise politicians?

Too many policymakers continue to confuse equality of results and equality of opportunity. Equality of results is incompatible with a dynamic society: it destroys the incentives that people need to work hard and produce new ideas. It is usually counter-productive; the egalitarian educational revolution of the 1960s ended up reinforcing social divisions. But equality of opportunity is the very essence of a dynamic society. It ensures that people are rewarded according to their contributions and promoted according to their merits.

Too many policymakers are locked in the age of big government. Big government has been failing since the 1960s in large part because a de-massified society needs de-massified solutions. Policymakers need to summon the resources of a wide range of institutions – local government as well as central government, NGOs and private-sector organisations as well as public-sector organisations, billionaire philanthropists as well as government ministers. Deloitte estimates that there are 650,000 social enterprises active in the United States, 62,000 in the UK and anywhere from 400,000 to 2.3 million in the EU as a whole. Governments also need to devolve as much decision-making to individual citizens as possible. A world where everyone has a computer in their pocket is very different from one in which computing power resided in a few government departments.

The smartphone revolution allows policymakers to tap into the wisdom of crowds to address the problem of inequality. Boston has a smartphone app that allows people to take photographs of potholes and graffiti and send them to the city council. The app automatically notes the GPS location. The same technique could be applied to schools that do not address dilapidated buildings or failing discipline, two problems that seriously demotivate students. The Defence Advanced Research Projects Agency has applied crowdsourcing to improve the design of military equipment. Perhaps the same approach could be used to improve the design of textbooks and school equipment. Jennifer Pahlka, the founder of Code for America, suggests that the best way for government to work better is not to make itself more like a private company but to make itself "more like the internet ... permissionless, open, and generative".

She might have added "scalable" to her list of virtues. One of the great virtues of modern technology is that it allows you to experiment with solutions and then scale up successful ones at breakneck speed (Reid Hoffman, the founder of LinkedIn, calls it "Blitzscaling"). When Salman Khan posted his videos on YouTube back in 2004 he was trying to tutor his own extended family. People outside his family soon began to watch the videos, including Bill Gates, who supported Khan with a grant, and the Khan Academy's videos have now been watched over 450 million times. These videos in turn contain 100,000 practice problems which have been solved 2 billion times.

The debate about the social impact of technology has been captured by purists. Techno-utopians argue that technological innovation will automatically bring abundance to all. Techno-pessimists argue that it will divide society into permanently hostile camps. Free-market purists argue that any interference with the digital dynamo will produce perverse effects. Statists think governments must have vast powers.

There is surely an argument for a digital version of the third way. We need to celebrate the power of technology but also recognise that it produces some losers, and to intervene to fix markets but also keep in mind that governments can be blunt instruments. Technology is providing us with the power to address the growing problem of inequality. Whether we can harness that power depends not just on the intelligence of technologists but also on the wisdom of politicians.

17

Work and the rise of the machines

Lynda Gratton

Technology is raising a host of questions about the future of work, but one thing is clear: successful organisations will have adaptability at their core

ACROSS THE WORLD, workers of every age are fascinated, perplexed and sometimes fearful about the impact that machines will have on work – and specifically on their jobs. They are not alone; government ministers worry about the loss of jobs and to what extent the "hollowing out of work" will extend beyond routine work. Will machines replace or augment jobs? Will the impact be felt in the next year, or play out across the coming decades? Will the jobs lost to automation be replaced, and if so, what will be the characteristics of these new jobs?

The advance of machine work is fascinating to watch: self-driving cars, Google's AlphaGo beating the world's best Go player, algorithms that evaluate job applicants or recommend a corporate strategy. Machines, it seems, can do almost anything that humans can.

I have watched the impact of machines on work over the seven years that I have directed the Future of Work Research Consortium (FoW). This brings together executives from more than 90 multinational companies from all over the world and from different sectors. Through workshops, focus groups and an annual survey we closely monitor the impact machines are having on work. It is challenging to keep pace with a constant stream of new developments in artificial intelligence (AI), big data, machine learning and a vast array of other technologies. With these trends arise new and competing schools of thought that interpret each new technological development and its potential impact on work and society. The resulting complexity means that right now it is impossible accurately to draw firm conclusions about exactly how each technology trend will affect work in the coming decades. However, there are emerging a number of broad questions that I believe reach into the heart of the debate.

Will machines provide time and space for critical thinking and focus?

The machines that are reshaping work are not neutral tools that emerge independently of individuals and society. Rather, their design and use depict how we choose to work as much as how these machines themselves influence work and society. This interrelationship between design and use is clearly seen in the development of the machines intended to save our valuable time. In the 1960s and 1970s domestic machines (washing machines, dryers, vacuum cleaners) created time for women who had previously engaged in domestic labour to join the workforce. In more recent times, the technologies embedded in the smartphone were designed to make life easier to connect to others, to get work done more effectively and to manage our personal lives more smoothly. The idea was that the time these technologies unlocked could be reassigned to focus on the valuable and uniquely human skills that require time and focus such as creativity, curiosity and innovation. There is no doubt that machines have indeed reduced the time it takes to do many previously human-capital-intensive tasks: analysts run algorithms to scan reams of data rather than manually searching for trends, and GPS enables those working in supply chains to track inventory with greater accuracy and less effort than ever before.

However, here is the paradox. Most machines are indeed there to provide the gift of time – and yet these technologies are the very same that are squeezing out the space for creativity and deep thinking, that invaluable human quality. The pervasive nature of technology, and in particular the rise of instant messaging and notifications, is resulting in profound technology overload. Workers are constantly interrupted (and indeed interrupt themselves) by a stream of information that distracts their attention and drains their cognitive resources. So rather than creating time for tranquil creative thoughts, our brains are busier than ever, assaulted with facts, pseudo-facts and rumour, all posing as information. Studies of our daily habits report that the average worker checks their mobile phone more than 150 times a day and is interrupted once every 10.5 minutes by instant messages, tweets and other notifications. After each of these interruptions it takes on average 23 minutes for those social-media users to get back to being focused.

So the question is this: is the future really 300 e-mails a day, and is the only way out of this a cleverer machine? It is clear that a growing challenge for technology and work is to find ways to reduce the amount of "noise" in the working environment and, crucially, to provide time and space for the most valuable of human qualities: creativity, judgment and decision-making.

Will machines make all the decisions?

One of the features that separate humans from other species is the capacity to make complex decisions. As machines become more sophisticated, is this role of decision-making as a uniquely human skill coming under increasing pressure? There is certainly evidence that in some circumstances machines can make better decisions than people. Already, for example, algorithms can predict events such as employee turnover and future job performance with more accuracy than managers can. An analysis of 17 studies on applicant evaluations concluded that equations outperform human applicant-selection decisions by 25%. So can we expect that decision-making and some aspects of managing people will increasingly be performed by algorithms?

Should we prepare for a future of work driven by predictive analytics? These are some of the questions explored at the World Economic Forum's meeting at Davos in 2014, where I chaired a debate under the motion "Computers will make better decisions than humans". It was clear from the debate that answering this question requires a deep understanding of how humans will work with, and relate to, machine learning. For example, will workers be persuaded to trust AI, and will entrusting machines to make decisions on their behalf run the risk of logical, but immoral actions? Can computers ever replicate values and emotions?

Studies show that many workers and organisations are reluctant to let machines make the final call. Whether diagnosing patients or forecasting political outcomes, people still consistently prefer human judgment – their own or someone else's – to algorithms. In part this is because, as neuroscience is showing, human decisions are more about emotions than programmable logic. It seems most people make decisions both cognitively and viscerally, drawing on the evolution of the pre-frontal cortex over the past 2 million years. This element of human development is not easily understood or replicated and cannot therefore be programmed into a computer. Put simply, the reality is much more complex than the simple prediction that machines will evolve from tools to decision-makers. But it also highlights the growing importance and value of the human skills of sophisticated judgment and decision-making.

Will machines shift the nexus of power from hierarchies to networks?

The emergence of inexpensive machines that facilitate communications, social networking and crowdsourcing has created opportunities to rapidly build networks of globally distributed people. These individuals and communities are able to connect across boundaries, share information at light-speed, swiftly attract new members and generate seemingly leaderless action. Will this increased peer-to-peer connectivity tip the axis of power from vertical hierarchies to horizontal networks? Will the business of the future be more like a "flatocracy", without hierarchy – operating through egalitarian power arrangements rather than by the power and orders of the leaders? When technology enables many people to have more information about themselves, their peers and the world, what role does a manager have, and indeed will leadership be transformed?

Interestingly, this potential levelling effect of technology has not yet materialised. Experiments to do away with hierarchical power structures have not been smooth. It seems that shifting the nexus of power in organisations from vertical to horizontal is proving tough: hierarchy and power seem highly resilient. An indicator of this is that since 1983 the number of managers employed in the US economy has nearly doubled. In part egalitarian power structures have failed because for most people their perception of status and overall standing in hierarchies is extremely important to them. Indeed, it is so important that it affects how they make decisions, how altruistic they are, and their overall mental and physical health. Machines might be capable of levelling the field; but it seems that the human preference currently remains wedded to hierarchy.

These questions are important and show that the potential impact of machines on work needs careful evaluation. However, it would be naive to conclude that the future of machines in the workplace will have a limited impact on work. Jobs, talent and practices are all under tension as AI becomes more sophisticated in data analysis and decision-making, as robotics displace routine work, and as machine learning revolutionises the work of even the most specialised and skilled workers. And while we may not be able to predict the future accurately, we can take away one important message: adaptability will become the defining aspect of successful organisations. This process of adaption has been at the heart of the research in the FoW consortium. There are four areas where we believe leaders must make significant changes to their work models:

- **Mend the broken career ladders.** The removal of swathes of middle-skill jobs (the "hollowing out of the middle") through automation has broken the historical career ladders. With middle-skill jobs gone, junior associates struggle to find the bridge to the next incremental step to the top. This trend is likely to continue and, while it may be hard to identify exactly which layer of jobs will be next to be disrupted, it is certainly possible to begin crafting non-linear routes to the top. Redefining what progression looks like in an organisation will be essential in addressing this challenge. Recognising that those middle-skill jobs are unlikely to return means it is imperative to embrace a more flexible approach to career development that may see people making lateral moves, or even leaving the organisation and rejoining it at a later stage. In the short term, the most important role for the talent-management function will be to coach and guide people through this transition and help identify valuable skill clusters and career-development opportunities.

- **Build engagement with ecosystems of talent.** The combination of powerful platforms for freelancers and reduced start-up costs for small, online businesses has created myriad working options for talented people. In 2014, for example, studies show that 53 million people in the US engaged in freelance work. This army of freelancers will only expand as the next generation enters the workforce. The result is that corporations with a single focus on full-time employees will miss out on some of the most talented people around. This will be particularly true for knowledge-rich companies where the product is ideas and creativity. Identifying where and how to engage talent on the periphery will become crucial, and will require a fundamental change in conventional recruitment approaches. Firms will have to develop a sophisticated understanding of both which people they want and what motivates them. They will also need a deeper understanding of the ecosystem around their organisation and the capacity to create relationships that go beyond the traditional employee–employer one.

  This means re-envisaging the employment relationship as a lifelong alliance. Talented people will use technological platforms to create value for themselves. They may be prepared to work within a company, but then want to leave to start their own business, returning later as a client or employee. So the emphasis will be on forging long-lasting relationships that outlive the formal employment term. In this "alliance model" both employees and employers want to add value to each other, in spite of the relative instability, lack of loyalty and long-term training investment in their coupling. For employees, this is an investment in the company's adaptability and value; for organisations, it is an investment in employees' employability and development.

- **Encourage lifelong learning.** Waves of technological disruption will destroy entire job categories and make certain skill sets redundant. As a consequence, the traditional model of front-loaded education followed by incremental skills development will be woefully inadequate. In its place will come an emphasis on lifelong learning that will in part be supported by an organisation, either through access to training, or via sabbaticals or flexible working to enable people to make the investment in boosting their productivity. There is no doubt that online, competency-based training and education will play a role. As technologies for learning advance, these will create opportunities for people to tailor a course that delivers exactly the training they need, at low or no cost, and at times that suit them. So in place of long, standardised degrees, this modular approach creates a personalised learning path. From a corporate perspective, this has the potential to deliver important data-driven insights on employees' preferences, engagement levels, learning styles and motivation.

- **Partner with machines.** Much has been written about machines destroying work, but for many workers the impact of machines will be to augment their work. For them, machines will be their partners in their knowledge work. This take on the future reframes the narrative from a challenge of managing job losses to an opportunity of augmenting and supporting employees' day-to-day activities. So the question becomes: what great feats can be accomplished by workers with their robotic co-workers? Where can workers collaborate with machines to achieve tasks they could not do on their own? What kinds of new jobs will these augmented possibilities create?

  The creation of these new augmented working models requires organisations to reimagine traditional job descriptions and creatively ponder how a job can be enhanced by innovative ways to strike a balance between humans and machines.

There is no doubt that the interface of work and technology will create ever more profound questions. Amid the rise of machine intelligence, where is the value in human work? What is the purpose of the technologies we are creating? Do we want machines to make our decisions for us? How do we want to live and work with each other?

These are more than just questions of profit or productivity. They are also questions of community, morality and values. In *Utopia*, Thomas

More wrote about an imagined island community. He described the inhabitants' practices of engagement and marriage, their celebrations of life and death, and their laws and ways of living. His was a much richer and more interesting view of the future than one centred only on technological determinism. Perhaps now, 500 years on from *Utopia*, it is time to reimagine our future in its totality and not let technological developments be the major framing.

18

Visiting hours: a short story

Alastair Reynolds

It's 2050, and technology raises questions of body and soul

CASSIE ETTINGER had always liked to climb. It had started with trees when she was small, then progressed to boulders and indoor training walls. In her teens she was already out-climbing adults twice her age. By the time she was hitting her 20s, struggling to motivate herself through a degree in architectural engineering, every available weekend was spent on rock. Each summer she pushed herself further than the last, scraping and begging her way onto climbing teams, slumming her way around the world in the last golden days of mass intercontinental travel. After she forged new routes up Rohlilahla and Bimbaluna, a measure of fame started coming her way. Sponsorship began to flow in from clothing and equipment suppliers. She got her face in the climbing magazines and the extreme-sports documentaries.

Then came the fall.

When the long rehabilitation was over, and she had begun to move on with the next phase of her life – the architectural career that had always played second fiddle to climbing – it was a fear of heights that stayed with her the longest. She had never really had it before the accident, except in the intellectual understanding that there was risk in being high, in trusting her life to fingertips and slender bits of metal. But she had never felt it in her guts, the way other people did.

It was different now.

As she climbed her way to the top of the stack, casting a critical eye over the work already done, the cuts and welds, the new windows and doors, the waiting stairwells and elevator shafts, each level gained brought an acute new appreciation for how far she was from the ground. She had to do it, though, just as she had to risk the occasional glance down to the dust and dirt of the construction site around the foundations. Her Geckoflex shoes and gloves gripped well on the curving alloy sections of the airframes. Cassie had Geckoflex kneepads as well, for the moments when she needed a little extra leverage. Geckoflex had been good to her in the sponsorship days.

Mostly, though, she tried to avoid using her knees.

The New Ofir Framestack was a mixed-utility construction project on the outskirts of Ofir, northern Portugal. Half residential, half business. From a distance, as it neared completion, it resembled a chrome scribble rising into the air, a dense cross-hatched grid of interlocking tubes. Or a pile of fat metal knitting needles, her critics said.

What they missed – or chose to miss – was that none of her Framestacks was random, or the product of haphazard design. Cassie took great pains in both conception and execution. Every angle, every slant, was critical. People had to live and work in these things, after all, and not get sick of them after a few months. The projects had to look good from all angles, inside and out, under all lighting conditions. Sunshine had to fall through their interstices in clean shafts, bringing illumination to gardens and vert-farms. Air had to circulate freely. Rain had to run off in effortless curtains, not accumulate in nooks and crannies. And the New Ofir Framestack had to endure a century's-worth of projected Atlantic hurricanes, with ample margins of error.

For all that the height bothered her, it was still good to get to the top. She pushed up, standing tall with her hands on her hips. For a moment or two she even felt she had to catch her breath. Her feet were planted either side of the curved ridgeline formed by the back of a Qatar Airways Boeing Dreamliner, the faint traces of its former ownership still showing along the sides.

Dreamliners were good. Carrie liked working with them.

In fact, there was another one coming in right now.

She had been watching the cargo airships as she climbed, wondering which was hers. They drank hydrogen from the sea-hazed towers of the offshore OTEC plants, filling their bellies with fuel. Airships were everywhere lately, but their schedules were slow and maddeningly prone to weather disruption. If you were working on a complex, time-critical project, you soon grew obsessed with the tiny droning specks of distant airships.

Cassie lifted a hand to her brow, visoring out the sun. The airship laboured in, its many ducted engines swivelling in accordance with some arcane attitude-control algorithm. Lowering down from its cargo gondola was the "new" airframe, already reduced to a pruned-back tube, with just the stumps showing where the wings and fin had once been mounted. There wasn't much "new" about it, of course. Twenty-five years of commercial air service, then the downturn, then the slump that never bottomed out, then the total grinding collapse of the entire global market for mass transport. But it was new to her. Not an airliner now, but a major structural sub-assembly. Most of the wiring would have already been stripped out, along with the seats, floors and luggage compartments. Gutted back to bare metal, there was a surprising amount of space in an airframe – and surprising strength.

Cassie waved the airship to come in closer. The airframe was already prepared for, a pair of welded saddles waiting to mate it to the existing stack. The airship began to pivot around to orientate the 'frame in the correct sense. Cassie wiggled her fingers and the winch spooled out, bringing the 'frame to within a few metres of the saddles. It loomed over her, ponderous and heavy where only a moment ago it had seemed airy and cloud-like.

Maintaining her balance, Cassie synched control of her tallest construction robot. The robot was a striding orange behemoth, somewhere between a tower crane and a Dali giraffe. Its neck rose high over the stack. The robot attached its own lines to the Dreamliner, allowing the airship to begin slackening off. Still synched to the robot, Cassie worked her way along to the saddle. The 'frame lowered down, wind whistling around its curves and singing in the winch lines. The final attachment, 'frame to saddle, was always the tricky part.

A face loomed in Cassie's upper-right visual field.

Incoming call, said the accompanying text. *Doctor Martin Abbate*.

"Not a good time," Cassie mouthed to herself.

But Doctor Abbate was nothing if not persistent, and there had to be a limit to how many times she could keep giving him the slip. It wasn't even the case that she disliked him, just that he was pressing her on something she would sooner keep pushing to the back of her mind. Not a bad man at all, and yes, he did sometimes have a point ...

She bit down on her irritation and took the call. Better to get him off her back now than have him pestering her for the rest of the week.

"I'm afraid I can only give you a moment, Martin."

"That's all I need, Cassie. You've been a very hard woman to track down lately. Work is going well, I take it?"

"Never better."

"That's good. I am so pleased that you have thrown yourself into this vocation. It has been very good for you."

The airframe's shadow deepened across her. "I said only a moment, Martin. I don't mean to be rude, but ..."

"That's all right, Cassie – I understand. But I must insist that you find time in your busy schedule for a visit to the clinic. At the earliest opportunity, too. There's been a development – a very significant one – and it affects the terms of the provision of care. I think it would be of great interest to you ..."

"I don't need to see her again," Cassie said. "Not now, not ever."

"I would request that you reconsider," Doctor Abbate said, with the mildest touch of sternness. "And sooner rather than later."

It was a million-to-one chance that one of the winch lines would sever at that precise moment, but that was what they had said about the rope, the one that had failed on her at precisely the moment she needed it most. The line snapped. The crane buckled back and the airship pitched. In the same instant the Dreamliner descended sharply at its tail-end, angling down like one end of a guillotine. A cold assessment of her predicament told Cassie that she had precisely no chance of avoiding its fall.

But she still flinched as it came down, raising a protective arm to shield her face.

It was true what they said. Old habits died hard.

*

Doctor Martin Abbate had a permanent stoop and a friendly, liver-spotted face. He had a tonsure of white hair and a pair of spectacles that she had never once seen him move from their fixed position above his forehead, where they studied her like an extra pair of eyes. She had known him since the accident and he had seemed old even at the start of their acquaintance.

"I am glad that you finally agreed to my request," he told Cassie. "Even if it took you a little longer than I might have hoped."

"I had work to finish in Portugal," she said. "We were already running a little late on topping-out and I didn't want to give my client any more ammunition."

"But all was well, in the end?"

She thought of the crashing Dreamliner, the sorry state of its buckled fuselage, the frustration of having to restart the procurement cycle again, bidding on another 'frame, waiting for another airship to haul it in. "Nothing we weren't used to," she said.

"You like this work, I see."

"There are clouds on the horizon," she admitted. "You wouldn't believe it, but the cost of those old airframes is starting to go up again. All of a sudden people are waking up to the fact that there's some value in them."

"We had Peak Oil," Doctor Abbate said, with a nostalgic smile. "You have Peak Airframe. But I wouldn't worry too much. You are very adaptable, I think."

"I'd better be. Look, I'm sorry to sound impatient, but there's still a lot I need to do ..."

He beckoned at the doorway she had been dreading. "Go through, Cassie. She's where she's always been."

She lingered before going in. The receptors in her nose registered the smell of disinfectant and powerful cleaning solutions. The photon-counting arrays in her plastic eyes registered sunlight coming through half-angled blinds, throwing a neat pattern of stripes across primrose walls, across racks of life-support machines and clean sheets, across the unmoving woman resting on the bed with her own blank eyes to the ceiling.

"Is there a problem?" Cassie asked.

"Not exactly. An opportunity, I would call it." Doctor Abbate beckoned her to join him by the motionless, bed-bound form. "The field of motor neuroprosthetics has come a long way in ten years. When we put your subcranial array in, it demanded difficult, invasive microsurgery. The risk was worth it, though, and we opened a channel into your functioning mind. We were able to ask you questions, to establish your degree of consciousness, your recall of past events, whether or not you were free of pain. It was a great comfort to your loved ones, to know that you were not in distress."

She recalled the grey texture of that awakening. It had been a regression to a child's sense of time, with endless minutes, endless hours, but no sense of days or weeks. Gradually there had come a kind of surfacing, rising back into something like normal alertness from a sea of confusion. Even now she had no recollection of the accident itself, or of the days leading up to it.

She hadn't been in distress, no. Not in physical pain, at least. But slowly she had gained an understanding of her situation, and that had been at least as traumatic as any discomfort. Completely paralysed, completely reliant on machines and around-the-clock medical care. Her spine shattered, her nervous system mangled beyond repair.

No kind of existence at all for someone who had lived to push herself to the brink.

"There was despair, to start with," Doctor Abbate went on. "But gradually you understood that there was still hope. The array was tuning itself, learning the secret language of your brain. Two-way communication was the first step. Then movement. Not of your own body, but of remote devices. A robot arm, to begin with. You fed yourself. Then a full-body total-immersion exo. What would have been prohibitively expensive a decade or two earlier was now affordable. Good for the telepresence companies, not so good for the makers of airliners! I still remember the day you were able to walk out of this room." His smile hardened. "But you didn't, not really. You controlled a walking, talking robot, but your mind was still in that body, still lying on that bed."

"Is there a point to this, Martin?"

He reached to the bedside and came up with a tablet. Through its translucency Cassie watched as his fingers tapped across a bank of glowing controls. "I said that what was difficult is now routine. The new arrays don't involve any kind of surgery. They grow in place from a microscopic subdermal seed, programmed to embed itself and establish the necessary neural connections. Self-learning, self-organising. How else are tens of millions of people able to hop from one remote body to another, as if it were commonplace?"

"And?"

"Raise your right arm, please."

She made to question him, but his tone had been so effortlessly commanding that she did it anyway, elevating her arm until her hand was level with her elbow. Silently, in exact synchrony, the figure on the bed raised her own right arm. The hand remained limp, but the gesture was an uncanny echo of the motion Cassie had just made.

"The array is detecting activity in your right procedural gyrus," Doctor Abbate said. "A few weeks ago we implanted a self-replicating electromyographic network into the arm, bypassing your own damaged nervous system. The network has grown and adapted. It's very responsive, isn't it?"

Cassie lowered her exo arm, and observed the figure on the bed mirror the movement. There seemed no lag between one and the other, no sense that one was the master, one the puppet.

"Why are you showing me this?"

"Because the arm need only be the start of it. The electromyographic network can restore full-body movement. You could rise from that bed and walk out of here, and you would feel as fully present in your real body as you now feel in the exo."

Cassie flushed under the scrutiny of Doctor Abbate, the weight of his well-meant expectations. She knew that she was already in that body; that in ten years no real part of her had ever left it. She was a paralysed woman confined to a single room in a private medical clinic that offered the best care that money could buy.

"Ten years ago," she started to say.

"Yes?"

"If you'd offered me this, the chance to walk again ... to be myself, fully back in my own body. I'd have wept for joy."

The faintest inkling of disquiet crossed his face. "And now?"

"She isn't me. I'm somewhere else. I can't bear the idea of going back inside her."

"You'd adjust, given time."

Her tone was purposefully cold. "I'm sure I would. But I don't want to." Speaking to him harshly, this kind man who had worked so selflessly, this man who only wanted the best for her, drew a cold line of sweat down her back. Of course there was nothing real about the sweat. Her exo body was an expensive, custom-engineered neuroprosthetic robot, made of plastic and alloys and warm, yielding composites. Parts of it were more like flesh than flesh. It had hair and pores but the one thing it did not run to was a sweat response. But her brain believed it should, and it had supplied the appropriate illusion.

"I died yesterday," Cassie continued. "There was another accident, just when you called me. The airframe came loose, and it crushed me. Destroyed my exo beyond any repair. For a few minutes I didn't have a body to go back into. They can't always assign you one at short notice, you see, no matter how good your plan is, but mostly we don't notice because we only make the switch when it's scheduled. But it wasn't scheduled, and I remembered what it was like to be back in her, back in that body."

Doctor Abbate's jaw tightened. "But it wouldn't be the same, not when you can move, walk ... you'd be free again. Free to live and breathe, to feel the sun on your skin ..."

"Free to be hurt," she answered, still with that same steely indifference to his feelings. "Free to be injured, free to be killed."

"You wouldn't be giving anything up," he persisted, his eyes pleading, like someone trying to foist a gift onto an unwilling recipient. "You'd still be able to ride an exo, just as you are now."

"But I'd know. In ten years I've forgotten what I'm really made of, how easily I can be damaged. And I don't want to be reminded of it." She smiled, trying to soften the blow of her words, to make him realise that she appreciated all he had done, all he meant to do, but that he was wrong.

"At least give the matter some consideration."

"I'm sorry, Martin. You're a good man. A good, kind doctor. But there's nothing to consider." She turned from the body on the bed, hating her own callousness, but knowing they would both be better for it. "Even if the flesh is willing, I'm not."

*

Later, when they were close to completion, the replacement airframe lowered into place, she found a ragged part of herself jammed in a crevice near one of the welded saddles. It was a hand and forearm, ripped off at the elbow, something they had missed during the initial clean-up.

She gave it a good kick, then watched as it tumbled down to the dust and dirt of the foundations.

19

Ma Ganga: a short story

Nancy Kress

It's 2050, a time of planetary peril and scientific activism

AT DAWN A YOUNG MAN stands at the top of a ghat. The broad, shallow steps lead down to a river stinking of decay. He is pale, freckled, overdressed for the heat in long trousers and thick shoes, quintessentially American. He clutches to his chest a foldable plastic cooler, the kind meant to hold a six-pack of beer, its black canvas straps dangling. He looks terrified.

Across the Ganges, the sun rises.

A group of barefoot old men in dhotis flow around him and wade into the water, past floating garbage, an armada of dead flowers, the carcass of an animal caught on some projection below the surface. The men's lips chant in prayer. They raise their faces, exultant, to the sun. A woman, graceful in a yellow sari, stoops to fill a vial with the holy water of Ma Ganga. Far out, a funeral boat cleaves through more drifting garlands.

The American descends the steps. He unzips the plastic cooler, as exotic in this place as a polar bear, and removes something from it. Then he squeezes shut his eyes and grimaces.

He cannot do it. It is not his to do. Not his river, not his country, not his choice.

He folds the top back over the cooler and climbs the wet steps.

*

"Dr Sanders! We are so delighted at your arrival!"

Seth Sanders shrank back, then tried to smile at the horde of people in the marble foyer of Global Enterprises Partnership. Horde? Well, no, seven people were probably not a horde. Diya would accuse him of unsociability.

But, then, when did she not?

"I'm glad to be here," Seth said awkwardly, and remembered to hold out his hand. "Dr Anand, Dr Müller, and ... uh ..."

"Nigel Harrington," said the tall man with the expensive suit, British accent and affronted manner.

"The division head for GEP," Dr Anand said merrily, "and our boss!"

Seth recognised her attempt to cover his gaffe, but he didn't know what to do about it. Harrington, the only non-scientist, had something to do with funding. Or maybe politics. Seth said, "Hi."

"Welcome to India," Harrington said frostily. "I hope we shall do great things together."

They all looked expectantly at Seth, the one who was supposed to do great things. If only Diya were with him! She always knew what to say, how to charm everyone. But she had gone straight to the hotel. And lately, how much was she willing to help him anyway?

The group still gazed at him. Seven pairs of eyes – blue, brown, grey – in seven faces carefully chosen for a balance of ethnicities, genders, religions. A press drone hovered overhead, relentlessly transmitting this photogenic array of optimistic scientific progress through private enterprise.

Finally Seth said, "Can I see the river?"

*

GEP had begun as big pharma, creating the weight-loss drug that had, for those who could afford it, completely turned off hunger with no side effects whatsoever, through genetically modified organisms that took up residence in the gut biome and sent precise signals up the vagus nerve. Within six months of its introduction, obesity had disappeared from middle-class America, black-market "imitators" scammed the gullible, and GEP became the fourth-richest corporation in a country increasingly divided between those who could or could not buffer themselves against runaway climate change, growing riots and the dangers of personal fat.

It was the riots that had prompted GEP to start, with much fanfare, its ecological division. They could not tackle climate change – apparently no one could tackle climate change – so they picked a smaller target, cleaning up polluted rivers. They then picked Seth, who had already done that, bringing him fame he did not want.

His field was epistasis, the effects of genetic mutations which depend on other mutations to function, thereby laying the groundwork for changes in protein function. Some changes were radical, even improbable. Most depended on new ways for proteins to fold. Seth had, through years of patient work, uncovered eight epistatic co-dependencies among the various mutations in different strains of his chosen bacterium. He then spent more years, long solitary good years, modifying the bacterium, always with a single goal in mind.

Then he reached it.

His genetically engineered bacterium destroyed a toxin discharged from textile factories into a small Indonesian river. Simpler genemods already processed the raw sewage discharged into the river, but nothing had been able to dismantle the industrial waste. Testing and retesting his bacterium in heat so great he could only work at dawn and dusk, Seth had been exultant. Ninety-eight per cent of the toxin was destroyed.

He hadn't known, hadn't imagined, what would follow. He'd expected journal publication and modest respect from his peers. Instead he got media celebrity. Robocams followed him, journalists besieged him, desperate people in poor countries wrote him heart-rending letters, detailing the deaths of children from river pollution and begging for help. Seth hated it all. He fled to GEP, accepting their offer to further refine his bacterium to handle other toxins. Seth held the patent. He would work in state-of-the-art Boston labs. He didn't even read his contract all the way through.

They "loaned" him to India. GEP's right to do that was in the contract. It made for great PR.

*

One of the women from the meet-and-greet, plus two bodyguards, led him to the Ganges. Which one was she? He couldn't remember. In her 50s, she looked like someone's grandmother. That should have made her easier to talk to, but Seth couldn't think of anything to say. He followed her on foot; evidently the river wasn't far away, which would be convenient for testing.

"GEP is building a direct and shielded walkway to a ghat," she said, "but it is still under construction. I am Saanvi Parth, by the way, University of Delhi. I think you don't remember."

"I'm sorry, I ..."

She threw him a mischievous smile. "It is all right, Dr Sanders. You were introduced to many people at once."

"Yes."

"And you don't like it. But we will now work uninterrupted. I am your partner in the lab, you know. Senior geneticist. Are you all right with the heat?"

"Yes. No. If we can just stop a moment ..."

"Of course." She handed him a water bottle. "Sit on that wall in the shade. It has never been this bad before. But you know all the effects of global warming."

He didn't, not by the numbers. His had been a life of air-conditioned labs, until Indonesia. He felt like a wimp. But her smile was forgiving.

"Your English is so good," he said, and hoped that wasn't a wrong thing to say.

"I did my graduate work and post-doc at Oxford."

They had stopped in a small square, or something like a square, lined with buildings and market stalls. Seth tried to sort out the whirl of colour and sound and odour. Temples, monkeys, women in saris, men in dhotis, motor bikes, dogs, police cams, beggars, a chanting group of men bearing a corpse towards the river, fruit and frying food and – yes – a cow wandering loose ...

All of it fascinating. All of it disconnected from him, ignoring him, letting him be a pure observer.

In that moment, bottle of warm water in hand and heat enfolding him, his heart turned over. Child of frozen Minnesota winters, of iron-range flintiness, of silent and dour parents, Seth fell in love with India.

*

He had met Diya at a party that GEP had made everyone attend. Seth stood in a corner with a glass of red wine he had no intention of drinking, wanting to be back in his lab, peeking at his watch. He made stumbling conversation. After half an hour – enough time already! – he darted towards a door, bumped into a woman and spilled his wine on her dress.

Two people standing nearby gasped.

"I'm sorry ... I didn't mean ... I ... let me pay for dry cleaning that!"

More gasps – why? Then the dress registered with him: it was made of leaves. Real, actual leaves, but something had been done to them: they rustled softly, barely above hearing level; they emitted a fragrance like pine needles; they changed subtly from gold to orange to pink – except where his wine had suddenly made them droop and go brown, like ... well, like dead leaves.

The wearer turned towards him, angry, and Seth nearly gasped. How could someone be that beautiful? Tiny, with masses of dark hair, green eyes, and flawless skin the colour of polished oak. She said, "Dry cleaning?"

"Or ... whatever ... I don't know ..."

"Clearly." Her anger changed to amusement. "You are Seth Sanders. The new boy genius at GEP. I hear wonderful things about you."

"I ..." He what? He was an idiot, he should stay in his lab, he didn't belong around people, especially people like this.

Someone came to his rescue. "This is Diya, Dr Sanders," the Someone said. "Diya Sodhi. The famous technofashion designer, you know."

He didn't know. She held out her hand. When he failed, in his befuddlement and embarrassment, to take it, she reached out and picked up his hand from where it hung limply at his side.

"Hello, Seth," she said. "I am glad to meet you."

*

Kanpur's tanneries, mostly Muslim-owned, were located in Jajmau, a Muslim neighbourhood. "There have been threats," Saanvi said, "and worse than threats. Some violence. When times are hard, people look for scapegoats."

Seth, studying the Ganges, barely heard her. Open drains carried tannery waste directly into the Ganges. The water was blue with chromates. Through an open gate he saw a yard with buffalo hides stretched to dry in the sun, a boy stripped to the waist treading hides in a vat of blue, geese and goats and clouds of flies. The air smelled like rotting carcasses mixed with battery acid.

"It has not changed here in thirty years," Saanvi said. "Chromium trivalent treatment gets hides more supple than the older vegetable-based processing. Too much of the CIII gets oxidised and turns hexavalent. A treatment plant was supposed to be built here, but corruption in Uttar Pradesh is worse even than in the rest of India and nothing was built. Everything has failed – national clean-up programmes, lawsuits, PR programmes, fines, sensors to detect violations. Kanpur has 406 registered tanneries plus more illegal ones, and standards are not enforced. Chromium pollution here is eighty times higher than legal limits."

"Eighty?"

"Yes. Ma Ganga weeps."

Seth looked at her then. "You really care."

For a long moment she didn't answer, and he thought he'd said the wrong thing yet again, that he'd been condescending. Finally she said, "Very much. I can do nothing about the biggest thing destroying India. Coastal flooding, crop failure, heat deaths ... But perhaps I – with you – can do something about this. And others may do something about global warming."

Seth frowned. "Who? What? Efforts by every single government have collapsed." From corruption, from influence by fossil-burning manufacturing, from faint-heartedness about geo-engineering, from inertia.

Saanvi didn't answer him. "Let us go back to the lab and get to work."

*

Diya, accustomed to fame, had been able to show him how to escape it, to elude for an hour or a day or even a week the tracking drones and holo reporters and garden-variety nuts sending him death threats or crank notes about "secret scientific discoveries". She took him to lovely places where no one cared about either microbiology or fashion: a cabin in the Alps, a ranch in Wyoming, a mountain in Nepal. Diya regarded these temporary vanishings from the mid-21st century as larks; Seth regarded them as oxygen. "You will get used to the attention," Diya said, but he never did. Between trips, he escaped to the lab, sleeping on a cot in his office at GEP.

"I am so proud of your work," Diya said.

And, later: "You work too much, Seth."

Later still: "You use your work to avoid real life, don't you? Including me."

"My work is my real life," he protested, knowing as soon as he said it that the words were wrong, but not knowing why. Her beautiful face froze into a mask; her musical voice took on edges sharp enough to cut diamond.

"Yes," she said. "For you, work is all that is real. I am only a dream you are tired of."

"Diya," he began, but then didn't know how to go on. His words would not come, his feelings as mis-folded as prions. All he could do was watch as she turned away, her black hair bobbing on her shoulders, her back rigid in a floating, cloud-like dress she had designed herself.

But she had come with him to India, although only on her own terms.

"I have seen Kanpur," she sniffed. "I don't like it. Crowded, dirty, poor. But my sister Ananya lives in the International District. I will stay with Ananya." And she had, in the India of air conditioning, smart-wired villas draped in bougainvillea, glittering restaurants and shops, all kept pristine by armed guards. As Seth and Saanvi and their staff worked longer and longer days, he saw Diya less and less.

Sometimes, he wondered if she noticed.

*

Weeks of gene editing, amplifying, testing, repeat repeat repeat. Weeks of genetically prodding proteins to fold differently, to become themselves but different. Weeks of steaming heat by the river and AC chill in the lab, of mind-straining work and computer analyses and sleepless nights.

And then, all at once, he and Saanvi found it. An altered micro-organism cleared a vat of river water of its chromates.

"I could drink that!" Seth exulted.

"Don't," Saanvi laughed.

"I need ... I'll be back in a few hours!"

He found Diya in her rented studio, surrounded by bolts of cloth, 3D printers, machines he did not understand. She bent over the work table, laser-cutting leather. "Diya! We did it!"

She pushed her goggles to the top of her head. "Did what?"

"The genemod! We found it!"

"That's great. Congratulations." She resumed cutting, without the goggles.

Seth went still. "That's not a buffalo hide."

"No. Calf. Wonderfully supple. I'm going to – "

"Where did you get it?"

Diya straightened. He saw in her eyes everything to come and that, in a strange way, she welcomed it. "The hide is local."

He could not stop himself, or what was unfolding between them. "From a cow killed here, against the law, and tanned with chromium salts poisoning the river."

"You're being righteous about pollution? You – an American? Your carbon emissions alone – "

"Diya ... the chromates end up in irrigation water, in vegetables, in milk, in breast milk – "

"You're lecturing me about India? You don't even hear yourself! You condescend to India by romanticising it!"

"And you do worse – you're a carpet bagger in your own country!"

Instantly he regretted his words. But maybe she didn't know what "carpet bagger" meant ...

She did. Her beautiful lips drew so tightly together that they almost disappeared. When she opened them again, she said, "This is not working, Seth. The marriage. It's not working."

The picture in his mind was a golden temple, crumbling to dust – he, who had never set foot in a temple in his entire life.

*

He didn't tell Saanvi, or anyone else, when Diya flew back to Boston. There were weeks of testing ahead before a formal report could be made to Nigel Harrington. The bacterium might mutate in the second generation, or the third, or the 26th. The crucial protein foldings might change. The genemod might react differently in different mixes of river water, in different concentrations of chromates, at different temperatures.

Seth slept on a cot in his office, while Saanvi went home to her husband, daughter and grandchild. She never commented.

The heat did not abate. Coastal flooding continued, displacing millions around the globe. CO_2 increased. Tropical diseases moved farther into temperate zones. Another climate summit failed. There were riots, coups, elections, including one in the United States. Seth ignored the news, until the night of gunfire close by.

Soldiers rushed to take up defensive positions inside and outside the GEP labs. But there was no need. They were not the target.

"Two hundred and seventeen dead," Saanvi said the next morning. "All in Jajmau."

"Why?" Seth said.

For the first time, she looked impatient with him. "I told you before. In bad times, people seek scapegoats. This attack ... most of the tanneries are Muslim-owned. There have been decades of simmering warfare between Hindus and Muslims, nationalists and globalists, corruption and reform, those with power and those who want power ... climate disaster just pushes everyone harder. Don't be such an innocent, Seth!"

"You're saying nothing can be done. About violence, about global warming, about ... anything."

"I'm saying governments are paralysed. But – " She stopped.

"Tell me. This isn't the first time you've hinted at ... something."

But all she said was, "We must get to work."

*

Four days later, GEP discontinued their project.

"I'm sorry, Dr Sanders," said the representative from Harrington's office. No grand committee this time. "HR will contact you about your departure arrangements. Right now, Security will escort you out. Here is a box for your personal belongings."

"But why?"

"GEP has determined that its Ecological Division can better meet its environmental goals through other projects."

"That's not an answer! And we're almost there!"

"I'm sorry."

Saanvi appeared from the lab, carrying her own box. "Seth, come."

"But – "

"Come."

She took him to a café smelling of turmeric and cumin. He could not force down his tea. Saanvi said, "Listen to me. There will be more violence. Rumours have been planted that the Muslims are planning 'to put even more poisons into Ma Ganga'. The bribes that GEP must pay the government have risen too high. The minister of environment retired and has been replaced."

"But what does all that have to do with science?"

"Drink your tea."

"I don't want the fucking tea!" And then, "I'm sorry."

In the gloom of the café, Saanvi's eyes were dark pools.

He blinked back tears. "It's just that we ... I wanted ... and it's all so hopeless. Everything, the whole planet."

"No. It is not hopeless." Her face hardened and she leaned towards him. "You are looking in the wrong place."

"What?"

"I will tell you something. It is not governments that will help the planet. It is not corporations like GEP. It is not universities. They are all too folded into the catastrophe, from self-interest. The only help will come from determined individuals."

He laughed. "Right. Like individuals can do anything."

"They can. I know some who will. They plan a co-ordinated geo-engineering project to inject a veil of genemod aerosols into the stratosphere, to deflect sunlight and cool the Earth."

Seth gaped at her. She was serious.

"We've known for decades that such a thing was possible – you must have known, too."

"Yes," he said. "But, Saanvi, no country has ever agreed to – "

"Not countries. I told you. This is privately funded by a billionaire who believes in it. Next month 20 planes will take off around the globe and inject the aerosols high enough to keep them stable in the atmosphere."

"The planes will be shot down!"

"Some may. By that time, it will be too late. The aerosols will be released and the planet will begin to cool."

"You'll change the whole climate! Crops will fail and – "

"Some will. Some will grow better. Over time, there will be a net gain. Seth – think. The climate is already changing, and for the worse."

"But – "

"Don't make me sorry that I am trusting you with this information."

"You are trusting an untested, radical, unknown procedure!"

"Not me. But, yes, the people involved are trusting the unknown. Sometimes there is no other choice. Here – this properly belongs to you, not GEP. You created it."

From a hidden fold of her loose trousers, she withdrew a stoppered vial.

*

At the top of the ghat, the sunburned American halts. A woman walks towards him. Dressed as he has never seen her, in a blue-and-gold sari, she touches his arm lightly but says nothing.

"Saanvi," he says.

Her dark eyes, the wrinkles deeper around them, wait.

"This is not how science is supposed to work," he says. "I can't take the responsibility."

"If not you, then who?"

He is silent.

"Seth, *nirdosh ek*, a protein can fold many different ways, yes? But it cannot fold an infinite number of ways. Everything comes to its limit."

Diya.

Saanvi adds, unnecessarily, "Even innocence."

A long moment. The funeral boat on the river discharges its load of cremated ashes. A very old woman pounds clothes on the bottom steps of the ghat. Two tourists snap holos. Overhead, a drone sails slowly by. The river gleams gold in the sunrise. It seems to Seth that he can see the "wet blue" from deadly hexavalent chromium, although he knows he cannot.

He goes back down the ghat, unstoppers the vial, and bends to the holy river.

20

Concluding reflections: lessons from the Industrial Revolution

Oliver Morton

To take stock of this book's insights into technology's future, look at the key event in its past

FOR ALMOST ALL HUMAN HISTORY and prehistory, technology was constant from generation to generation. People used tools similar to those that their mothers and fathers had used, or that had actually been passed down to them; the same trowels, hoes, pestles, pots, needles, knives and the like. The world was not unchanging, and innovations mattered: the chimney changed the nature of the house, the stirrup the role of the horse. But change was slow, and technologies could be lost as well as gained. When Filippo Brunelleschi set about building a dome for the cathedral in Florence he found himself taking as his model, or at least inspiration, the dome of the Pantheon in Rome, dedicated to the honour of Agrippa. Over the intervening 1,300 years the art of dome building had been lost in western Europe.

For much of the world a lot of technology is still experienced in this inter-generationally stable way. For many purposes many people still use the technologies of their parents' generation. As David Edgerton, a historian of technology at King's College London, argues in *The Shock of the Old* (2006), a current fascination with technology-as-innovation hides the history of technology-as-used, and it is technology-as-used (often as basic as the rickshaw or the condom) that underpins many aspects of human life, defining what can be done and how hard it is to do it.

This stability is not just something that applies in rural settings and the developing world. Vaclav Smil of the University of Manitoba points out in *Prime Movers of Globalisation* (2010) that the two fundamental technologies on which today's trade and transport depend are long-established: the diesel engine dates back to the 1890s, the gas turbine to the 1930s. They took time to attain their dominance, but for most of the past century the diesel engines of ships, trains and lorries have moved most of the world's goods, with jets now providing for things in a hurry.

And as Smil argues elsewhere, perhaps the most important technology of the 20th century – and certainly one of the least appreciated[1] – has remained unchanged at heart since its inception. In the late 1900s and early 1910s Fritz Haber and his colleague Carl Bosch of BASF (a chemicals company) and the University of Freiburg developed the first practical way of "fixing" the inert nitrogen in the atmosphere into reactive forms of that element, forms from which artificial fertilisers and explosives could be made.

This allowed warfare on a scale never previously imagined: in the second world war, by one calculation, more than 6 million tonnes of high explosive were used, a cornucopia of destruction which would have been inconceivable without artificial nitrogen fixation. But the technology helped to feed far more people than it helped to kill. Artificial fertilisers based on industrially fixed nitrogen made it possible for the world to support a population four times larger at the end of the 20th century than it had been at the end of the 19th. Artificial fertilisers remain crucial to human civilisation; and though the scope and efficiency of their production have been greatly expanded and improved, as have the capabilities of diesels and turbines, they continue to be made by the Haber Bosch process.

There is a slow and solid technological substrate to the world, and there always has been; there are deep continuities that neophilia ignores and hype seeks to hide. At the same time, though, the worlds of Brunelleschi and Bosch were fundamentally different. The second half of the 18th century and the first half of the 19th saw closely coupled changes to the nature of technology and of society. First in Britain, then around the world, the idea that the next generation's life would be largely indistinguishable from the previous one's was overturned.

The Industrial Revolution changed not just the technologies with which things were made; it changed the rate at which technologies themselves changed, and the rate at which businesses built on those technologies sprang up and fell away. It produced a world of constantly growing technological potential and saw constantly growing economies built on the back of that potential. The world's gross product, in something close to a steady state throughout all previous millennia, started to grow exponentially. Though the exponent changes over time, it continues to do so today.

A will of its own?

It is common and peculiarly tempting to see the change as being due to technology itself, most notably that of the steam engine in its modern form; that when technology unleashed new power and thus new possibilities it also made them inevitable. The sense that innovation drives the dynamics of history as a piston drives a wheel can be found in all sorts of places. You see it in the way that innovation, rather than its use, dominates the way people tell stories about technological history, stressing the new because it is the new that seems to drive the story. You see it in economic theories in which technological change is the thing outside the system that accounts for its long-term dynamics, the residual source of growth inexplicable from within. You feel it in the vague but persistent sense that progress and its problems are accelerating willy-nilly. These ideas are not all of a piece; as Ryan Avent notes in his essay (Chapter 6), economic-growth figures do not reveal the sort of accelerating progress that people perceive in their everyday lives. But they are all shaped by a sense that technology has a certain autonomy – a will of its own.

To experience this idea at its optimistic full throttle, have a look at *What Technology Wants* (2010) by Kevin Kelly. Kelly, the founding executive editor of *Wired* magazine (and a friend and former colleague of mine), sees technology as an entity developing according to its own rules and its own logic, something not unlike a force of nature but more intentional than such forces are normally taken to be. His "Technium", as he terms the sum of all things technological, is one expression of a universal drive towards greater connectedness and greater complexity. Since Kelly thinks people benefit both materially and spiritually from connectedness and complexity all this is for the good: by understanding what technology wants and by helping it to achieve its aims humans will further their own ends, too. But this optimism depends on accepting the providential alignment of the Technium's interests and humanity's. The reader is left suspecting that if what technology wanted and what was best for people diverged, it would be the humans who had to realign their priorities. And that is, indeed, how many people experience the world.

Kelly is much more forthright (and cosmically ambitious) in the way he treats technology as a power unto itself than most; but similar if less clearly expressed ways of thinking can be found in any number of less ambitious writings about technology emanating from Silicon Valley and elsewhere. Marxists have a name for such thought: fetishism. As Karl Marx said in *Capital* (1867), the modern world obfuscates its underlying social structure – who does what for whom and why – by endowing inanimate objects with powers that they cannot possibly possess of themselves in a way that seems supernatural. He wrote:

> [In] the misty realm of religion … the products of the human brain appear as autonomous figures endowed with a life of their own, which enter into relations both with each other and with the human race. So it is in the world of commodities with the products of men's hands.

You do not have to be a Marxist to be struck by the applicability of such a critique to talk about technology as an autonomous entity that follows its own laws, reshaping the social setting it finds itself in. Such talk glosses over the question of who chooses the technologies, how their spread comes about through markets and by other means, and who benefits. This obfuscation helps the status quo. But you can be a critical Marxist and overlook the problems with accounts of technology as an actor in its own right – indeed you could, at times, be Karl Marx and do so. Marx's views on technology varied over time, but they often read as quite determinist, seeing technology in itself as a changer of the world – as when, in *The Poverty of Philosophy* (1847), he wrote: "The hand-mill gives you society with the feudal lord; the steam-mill, society with the industrial capitalist."

As Andreas Malm of Lund University argues in *Fossil Capital* (2016), a fascinating reappraisal of the Industrial Revolution, the relationship between society, the steam mill and the industrial capitalist was a lot more complicated than that. Steam power is often seen as fundamental to the Industrial Revolution; the key step that made it possible. But surveying various accounts of why steam took off in Britain in the late 18th and early 19th centuries, Malm finds them wanting. Britain was not uniquely adept at finding and burning coal; steam power was not in its early years a clearly superior way of doing things, nor was it more easily improved than the alternatives.

What is more, the industrial capitalist was already on the scene. The exponential growth in output that is commonly taken as symptomatic of the Industrial Revolution starts well before the steam engine gains currency. British cotton production took off in the 1780s thanks to the introduction of a range of mechanical improvements flowing from a culture that both encouraged innovators and systematically reinvested a significant fraction of innovation's rewards (rates of return often exceeded 30%) in measures that further lowered the costs of production. And this expansion was undertaken with a form of motive power that was well known when Agrippa paid for the Pantheon: water wheels.

To the extent that this is appreciated by those proposing broad-brush histories of technology or energy use it is normally seen as a transitional phase, with steam engines arriving as saviours around the point at which the inherently limited capacity of water wheels is being outgrown. But this is simply not true. In 1983 Robert Gordon published a detailed study of the watersheds of Britain's industrial north. By 1838, a point at which the Industrial Revolution was well under way, they were all being more thoroughly exploited than ever before. And yet in no river basin was as much as 10% of the hydropower potential being used. In the basin with the most potential, that of the Trent, industry was using less than 2% of what was available.

Steam was not necessary for exponential industrial growth; it was used not because there were no alternatives, nor because it was cheaper than those that there were. It was used because it suited those making the investments, for various reasons. It freed them from constraints of time and place, allowing them to put mills where they wished, and to alter the speed of their operation with ease. Thus it allowed industry to be highly concentrated, encouraging and making use of deeper pools of labour. The reasons owners and, to an extent, workers preferred steam to water were not to do with the inherent power of technology; they came out of the social relations between the two groups. Technology did not cause the revolution; it was the revolution's child.

Steaming ahead

Why does the early history of the steam engine matter today? For three reasons. First, it highlights the central role of capitalism. The reinvestment of capital in future market-driven growth creates a demand for technology completely unlike that seen in any previous mode of social organisation. This is not to say, as some would hold, that the purer the capitalism the better the technologies. In the 20th century governments played crucial roles in all sorts of technological development, and often in technological diffusion too. But people operating in modern market economies have shown a capacity to demand and deploy technological change unimagined under previous conditions.

A phenomenon often taken as evidence of the independent dynamics of technology illustrates this nicely. Vast numbers of innovative ideas, from the diesel engine to the telephone to the light bulb, turn out, on close inspection, to have occurred to many people at the same time. Since pioneering studies by Langdon Winner and Robert Merton in the 1970s, simultaneous discoveries and acts of invention have come to be recognised as the norm – representing, to someone like Kevin Kelly, evidence that there are things which technology "wants" to do next. It seems to me, though, that it simply represents that in capitalist systems there are high rewards for thinking of new things to do and a finite number of novelties to aim for. As evidence, consider the things which technology might have done in the pre-capitalist world, but chose not to. Buttons and front-wheeled wheelbarrows could have been invented in the classical world; they were not.

Second, the way that steam power came to be seen as causative for, not merely central to, the Industrial Revolution tells us something about what people want from technology. The steam engine became an icon for the idea that technology operates according to its own agenda because steam engines themselves seem to do precisely that. From their earliest days, the idea that steam engines work on their own is treated with a fetishistic, superstitious awe by both the technology's devotees and its doom-mongers. Both hope and fear have their role to play: the hope that technology will save us, the fear that it will damn us. Frankenstein, after all, is a story about an autonomous technology.

Computers play a similar role today to that played by steam engines among Victorians. They are the technology that stands for all technologies and, through their own apparent autonomy, instantiate ideas about the autonomy and agency of all such artifice. When they say no, or say yes, or "make" decisions, they always do so in ways that are determined by the programs that have been written by human hands, for human purposes, with human flaws. But they do so in ways that make it hard not to imagine the agency as residing in the device itself. The current move towards artificial intelligence (AI) is so fascinating precisely because it seems to push this perceived autonomy to its limit. *Pace* Kelly, I do not believe that technology wants anything; but I believe that people want technology to want things, at some level, and that some of them are working hard to make it do so.

This focus on autonomy is understandable, given that it is in information technologies that the most obvious innovative progress has been made over the past few decades (I say obvious because, as Robert Carlson reminds us in Chapter 3, the progress in biological technologies has been enormous too). There is a risk, though, that preconceptions about autonomy as the central technological issue of the age may skew the ways in which we look to the future. Take, for example, the future of weaponry. In his fascinating essay (Chapter 11) Ben Sutherland details many aspects of autonomy, or intelligence, that could be applied to weapons, from bullets that can turn round corners onwards. But he does not choose to explore the suggestion in Frank Wilczek's essay (Chapter 2) that far better control of nuclear physics might well quite soon allow "ultra-dense stores of energy that are smaller, better controlled and more versatile than today's reactors (or bombs)".

This devastating semi-aside puts me in mind of remarks made some years ago by Freeman Dyson, a great mathematical physicist and sometime colleague of Wilczek's. Dyson said that if there were a single physical truth he would not want to see discovered it would be a means of detonating fusion weapons (H-bombs) without fission weapons

(A-bombs). This, it seems to me, is one of the things that Wilczek is speculating about.

At the moment, cramming together hydrogen atoms powerfully enough to set off an H-bomb is possible only if you split apart atoms of uranium or plutonium in a chain reaction first; all fusion weapons have a fission weapon built in to get things going. Designing fission weapons is comparatively easy: there was a design already developed when the first Manhattan Project scientists reached Los Alamos in 1942. Amassing the right isotopes of plutonium and uranium in the quantities needed, however, is much more difficult. It required a fair fraction of the US's GDP to do so in the 1940s. It is still not something easily undertaken by any entity smaller than a state – or that can be done without other states noticing and taking an interest. The fact that the ability to make nuclear weapons has spread less far than was imagined in the 1950s and 1960s, and with less terrible results, is largely because amassing the fissile materials is hard, and sanctions and other forms of suasion – including pre-emptive strikes – can make it harder still.

But if Wilczek is right, and computer-augmented nuclear physics makes it possible to design new ways of getting energy out of nuclei in something like the way that chemistry has allowed the design of new sorts of molecule, this barrier could be removed. In such a world you can imagine nuclear weapons made with a far less obtrusive technological footprint than is required today, with more easily sourced materials and in great profusion. There could be suicide bombers with kiloton yields. This is a prospect that seems to me rather more worrying than the idea that AIs will rise in revolt and enslave us all. It is also one much less discussed.

Unintended consequences

The third reason to concentrate on the advent of steam is that it represents the biggest-ever example of unintended consequences. The idea that carbon dioxide played a role in keeping the Earth warmer than it would otherwise be was appreciated by some by the second half of the 19th century; but it was not until the middle of the 20th century that it became clear that the carbon dioxide given off by the vast profusion of fossil-fuelled devices that industrial capitalism had brought into being would probably change the climate, and not until the end of that century that this alarming prospect came to be taken seriously. Thus the decisions of 19th-century capitalists changed the planet in ways they never imagined.

The consequences of using the atmosphere as a rubbish dump could have been even more dramatic than they are turning out. As Paul Crutzen, a noted atmospheric chemist, has pointed out, the chemical pathways by which the chlorine in refrigerating gases destroys ozone in the stratosphere were not known at the time when those gases (chlorofluorocarbons or CFCs) were developed. So an ozone-destroying chemical was let loose in the stratosphere, with potentially profound consequences for the global environment. Happily, the problem proved tractable after it was noticed; though the ozone layer is still damaged, it is not worsening and probably on the mend. But if chlorine's chemical behaviour had been like that of the closely related element bromine – or if bromine had been used on the same industrial scale as chlorine in similar compounds – things would have been very different. When it comes to ozone, bromine is to chlorine as combine harvesters are to scythes. The ozone layer would not have developed a hole at the uninhabited end of the Earth half a century after the chemicals were introduced; it would have collapsed quickly and more or less completely over the entire planet.

It is remarkably hard to generalise about how to face this problem of unintended consequences while continuing to innovate, but one obvious point is that the problem tends to be largely one of ignorance (though a failure of the imagination is often also at play). Although this ignorance is sometimes wilful or even feigned – we should never forget the shameful response of the world's tobacco companies to the discovery that an unintended consequence of their nicotine-delivery technology was the death of millions – it can also be genuine. If the people who had developed CFCs had thought harder about what they were doing they would not have made better decisions, because the science necessary to understand the consequences of their actions was not available to them. They did not simply want for the facts. The ideas, concepts and tools with which to think about the problem were absent.

This is not, I think, something that is solvable. Being diligent about known unknowns when it comes to assessing the impact of a technology obviously makes regulatory sense. But there will always be unknown unknowns, too. Perhaps the most obvious area where they may lurk at the moment is in the operation of the mind, much of which is considerably less well understood than the finer points of stratospheric chemistry were in the 1930s. As technologies become more closely entangled in cognition – as they come to play a role, as our essayists have imagined, in the augmentation of memory and the senses, in interpersonal communication and in the workings of the imagination – the opportunity for them to effect lasting unexpected change probably grows. People with mental prosthetics and enhancements will think differently as a result. It is by no means clear that such changes will be harmful. But it is clear that we can't fully anticipate them at present.

Possible responses

When unexpected consequences do crop up, though, we can be sure of two things. One is that the problems will spur the development of new technologies intended either to replace the technology which is causing the problem or to attack the problem directly in a way that puts things right. For CFCs, the replacement route was taken, with new technologies designed to do the job of the old ones. Something similar is being seen in climate change. While some, including Malm, would like to remake the capitalist social relations that underlie the carbon-climate crisis – "Capitalism vs the Climate", runs the subtitle of Naomi Klein's bestselling book *This Changes Everything* (2014) – most efforts are going into the replacement of energy-inefficient technologies with energy-efficient ones, and on the replacement of generators that liberate fossil carbon with those that do not.

There is, though, the possibility of not just replacing the old technologies with new ones that do the same thing, but also supplementing them with technologies designed specifically to address the problems. In the case of climate change, this might involve geoengineering technologies that could brighten the Earth's clouds, or use particles in the stratosphere to scatter sunlight, or pull carbon dioxide out of the atmosphere. It is more or less impossible for ambitious climate goals, such as limiting the amount of warming to 1.5°C, to be assured of success without either some sort of geoengineering efforts along these lines, or the use of time machines that would allow policy changes to take effect in the past rather than the future. As Wilczek points out, time travel is one of the science-fictional technologies that really is ruled out by the laws of physics; geoengineering technologies, however, while sharing a science-fictional glamour, are plausible, even if both their efficacy and the risks their side effects might pose are still uncertain. This means either that talk of strictly limiting the temperature rise should be much more guarded than it is, or that the role of geoengineering should be more explicitly discussed.

Such discussion should highlight the immense effort that would be needed to develop these currently more-or-less imaginary technologies into something that could be deployed in a safe, just and governable way. There is no precedent for the responsible and inclusive development of such an ambitious technology on a global scale. But developing the technology without wide consultation and explicit discussion of its potential to do damage would be a terrible mistake.

One of the visceral reasons people worry about geoengineering is that they want something more enduring, or fundamental, than a "technological fix". One can sympathise. But at the same time, one should understand that all technologies are fixes. They meet needs, and they create needs. The idea that geoengineering might solve the climate problem once and for all is absurd; but so is the idea that any technology, or for that matter any institution, can be fixed and permanent in a world where capitalism drives perpetual growth and perpetual change.

This brings us to the second thing we can be sure of when it comes to any response to future unexpected consequences: those responses will, themselves, have unexpected consequences, and innovators will explore ways to mitigate those consequences that will have unintended consequences themselves. There is no obvious way off this ladder, or treadmill. As W. Brian Arthur, an economist, wrote in *The Nature of Technology* (2009): "Problems are the answers to solutions."

Technology can never be relied on to solve problems in the absence of social action; one of the dangers of fetishising technology as an actor in its own right is that it obscures this point. Good solutions will rarely, if ever, be implemented through technology alone. And technology will never be the last word on anything; there will always be something new to try, some other galling thing to seek to set right. The centuries of ceaseless technological change are not going to come to an end; they may only just be getting going. A clear understanding that technology does not have its own agenda but serves the agenda of others, and that it necessarily creates new needs almost as effectively as it meets old ones, will make this change easier to navigate responsibly. But it will never bring the changes to a halt.

Note

1   When I once mentioned its centrality to the great economist Robert Gordon, whose work on the peak and decline of innovation Ryan Avent discusses in Chapter 6 and whose work on water wheels features in this essay, he looked at me as though I were deranged.

Acknowledgements

NEVER PROPHESY, so the saying goes, especially about the future. I am grateful to all the contributors to this book for so gamely ignoring this and being prepared to cast their minds three decades ahead.

One of the joys of working at *The Economist* is that it takes only a short walk down the corridor to find the best possible advice. Many colleagues chipped in with valuable suggestions. Oliver Morton and Tom Standage, in particular, were hugely helpful in shaping the outline and recommending writers. Matthew Symonds and Natasha Loder gave thoughtful feedback. Zanny Minton Beddoes encouraged me to do the book, and Patsy Dryden helped to make sure I could keep on top of the day job.

Clare Grist Taylor and Ed Lake of Profile Books were enthusiastic supporters of the project throughout, and remarkably patient. Paul Forty calmly shepherded things through. The excellent Penny Williams, David Griffiths and Pip Wroe helped with, respectively, copy-editing, fact-checking and charts.

Megatech looks to 2050, but it took chunks out of 2015 and 2016. Special thanks to my wife, Gaby, for her megatolerance of the time taken up by all the future-gazing.

Daniel Franklin

Index

Page numbers in *italics* refer specifically to Figures.